THE PALEO PERSPECTIVE

THE PLIGHT OF PREHISTORIC MAN IN MODERN TIMES

BOB DECO

ISBN: 978-1-4834-8817-2 (sc)
ISBN: 978-1-4834-8816-5 (e)

Library of Congress Control Number: 2018908059

Lulu Publishing Services rev. date: 02/04/2019

Special thanks to Janice Stanzin and Sara Ging for their help with my grammar and spelling, and Bobby Petraglia for illustrations of man's ancestors.

CONTENTS

*"We are just an advanced breed of monkeys
on a minor planet of a very average star.
But we can understand the Universe.
That makes us something very special."*
- Stephan Hawking

FOREWORD

"It is far better to grasp the universe as it really is than to persist
in delusion, however satisfying and reassuring." - Carl Sagan
(American author, astronomer, cosmologist, and astrophysicist)

In the 1984 movie *The Terminator*, an artificial intelligence system
(computer network) called Skynet suddenly becomes aware of its own
existence—it becomes self-conscious. One might say that the machine
became awake. Likewise, as man's intelligence evolved, he too became
more self-aware—more awake. When asked to describe himself, the
Buddha is credited with describing himself not as a god, not as a prophet,
but as awake. What does it mean for a human being to be self-aware, to
be fully awake?

Part of being awake means being aware of the present and operating
in the present. But the present contains not only seeds for the future, but
also residue of the past. And so being fully self-conscious not only includes
consideration of what is to come, but also acknowledgement of the past's
contribution to the present. Sometimes we must look at the past to more
fully understand the present. Becoming fully self-aware is essential to
understanding the past, the present, and the future.

In *The Terminator*, the machine's self-awareness not only included
how it was created, but also its plans for the future—world domination,
of course. Being self-aware as a human being includes considering how
our present actions not only affect our own personal future, but also the
futures of the people around us and the future of the planet. It includes
thinking about how we as a species (Homo sapiens) got here, and what
part of our ancient past is still with us today. We cannot be fully awake as
an individual or as a species if we ignore our ancient past, our evolutionary

past. Human evolution is itself a journey toward greater self-awareness—dare we say a journey toward enlightenment. To be sure, we cannot be enlightened without first being awake.

Man, perhaps out of hubris, has had a tendency to separate himself from nature—from the animals and various other life forms that surround him. He has aspired to see himself above nature—even in charge of nature ("Let them have dominion over the fish of the sea, the birds of the air, and all the wild animals," Genesis 1:28, *The New American Bible Saint Joseph Edition*). We often find ourselves looking for that one special thing that separates us from the rest of the animal kingdom. If we think of ourselves as special ("God created man in his image," Genesis 1:27) it is easy to forget that we are subject to the same rules that govern all of nature. That is unfortunate because we can learn a lot about ourselves if we allow ourselves to be analyzed and scrutinized through the same lens that science looks through when it observes the rest of the natural universe. We mustn't fear what science can tell us about ourselves. True self-awareness is a blessing, not something to avoid. We should instead embrace what it truly means to be a human being.

Education is about being awake, whether that elucidation is formal or self-directed. Willful ignorance is a self-imposed stupor. One need not be a scientist or a professor to be informed. One can learn a lot about oneself and the world just by being open minded, introspective, and curious about human nature. It behooves us all to be life-long learners—to strive to become more awake.

This book is for the life-long learner, the curious, the seeker, and the skeptic—for those of us who want to know "how and why." It is for those who have a healthy respect for science and value its systematic study of our natural world. It might be less comfortable for those who get a little "rattled" if their long-held beliefs are challenged. And although this book does express a particular worldview, it does try to maintain a healthy level of respect for other various worldviews. At the very least, I hope that what follows is thought-provoking.

ABOUT THE AUTHOR

Let me start off by declaring that I have never before attempted to write a book. If you choose to read even a small portion of this book, you will soon discover that I am not a gifted writer. I confess that I have always struggled with my writing skills. To that end, I have labored to keep my sentences short and to the point.

My undergraduate degree was in engineering, and engineers are not noted for their literary skills. During my undergraduate engineering studies at Stony Brook University (New York) we used to joke: "When I started this program I couldn't even spell "enjineer," now I are one."

I have always been an inquisitive fellow and my lack of literary skills has not discouraged me from pursuing my academic curiosities. I have been involved with various continuing education classes at Stony Brook University for many years, and fancy myself to be a lifelong learner—a seeker at heart.

Although my formal training was in engineering and physical science, I have always been curious about human behavior. Why do we do the things we do? Why do people believe the things they do? As a youngster I attended Catholic masses with my parents and listened very intently to the biblical readings and the sermons. Religion, theology, and human nature have always fascinated me.

"Perplexity is the beginning of knowledge."
- Khalil Gibran (Lebanese-born American writer and poet)

When 9/11 came along my interest in religion piqued and I found myself reading a variety of books on religion and God—Islam in particular. Award-winning British author and commentator Karen Armstrong has written many books on this topic. Three of her books I found helpful were *Muhammad*, *Islam*, and *A History of God*. My readings on religion were not isolated to Islam or to Karen Armstrong. I also read books on Buddhism, Christianity, Atheism, and the anthropological origins of spirituality and religiosity. I wondered about man's propensity for religiosity in general. Was it cultural? Was it genetic?

Does God exist, or is God an invention of the human mind? Books such as *The God Gene* (Dean Hamer), *The God Delusion* (Richard Dawkins), and *The Belief Instinct* (Jesse Bering) wrestle with these very questions.

My interest in theology eventually led to a desire to better understand human nature per se. The scientist in me reasoned that one couldn't fully understand present day human nature without first considering how it is that we (Homo sapiens) came to be, and how evolution was responsible for our current physical and psychological design. And so for the last several years I have been attending classes and lectures on anything that has to do with archeology, anthropology, and evolution.

I found myself jotting down things that I had learned—thoughts that entered my mind. I kept a notepad by my bedside and a recorder in my car. I could be reading the newspaper, listening to the radio, or lying in bed and whenever one of these mini-epiphanies occurred, I wrote it down or recorded it.

Perhaps responding to some inner urge to communicate with my fellow man, I began to organize these notes, recordings, and thoughts for the purpose of assembling them into some sort of coherent, readable package. The culmination of these efforts is *The Paleo Perspective*.

ABOUT THE BOOK

"Never, ever use repetitive redundancies. Don't use no double negatives. Proofread carefully to see if you any words out." - William Safire (American author, journalist, and presidential speechwriter)

As previously alluded to, this is a book written by a layman. It is for the layman—for the general public. Much of what is presented is either fairly common knowledge or information that can be easily verified at the public library. It is not a science textbook on anthropology or evolution, and although it covers some topics that might be classified as sociobiology or evolutionary psychology, it is not an in-depth analysis of either.

That being said, *The Paleo Perspective* does represent a considerable amount of research—albeit research that anybody could do with a computer and the Internet at their fingertips. When I felt it was appropriate, I did provide the reader with the sources of my information. My references are imbedded in the text, at the point of reference, rather than at the end of a section or at the very end of the book. I did so because I wanted the reader to feel comfortable with the credibility of what was being presented, as it was being presented—no flipping pages for references or footnotes.

This book is about human nature. It is about everyday things such as greed, conflict, empathy, and cooperation. It's about how people behave and why they behave the way they do. *The Paleo Perspective* deals with morality, religion, economics, and politics—all those topics of discussion that can get you in trouble at a family holiday gathering or at a dinner party with friends. Ultimately, it is about a particular worldview—a particular mindset—one that I suspect many other people share.

Admittedly, many of the ideas and concepts presented in this book are not original or unique, and are in fact well documented and part of a large body of commonly accepted scientific knowledge. What I do think is unique about this book is the breadth of the application of these particular ideas and concepts to the world around us—to our everyday lives. Of particular value is the piecing together of these ideas and concepts to form a coherent perspective—a perspective that I am calling *The Paleo Perspective.*

Book Organization

Because a *paleo perspective* is a scientific perspective, and because the theory of evolution is central to the theme of this book, there will be several brief chapters devoted to covering some basic scientific concepts that I feel are essential for the understanding and appreciation of the book. All discussions will be succinct and rudimentary—no graduate degree necessary. I make no assumptions about the reader's scientific background, or lack thereof.

Part I of the book will start with a basic timeline that takes us from the big bang to the present. Significant milestones in the evolution of our universe and life on Earth will be highlighted. There will be brief and concise discussions of Darwin's theory of evolution, basic genetics, and the human brain. These topics are extremely pertinent in any attempt to view human nature through a scientific lens.

Part II will focus on human nature as it pertains to behavior on an individual basis, and as it pertains to human behavior within a group setting. To that end, there will be a segment on man's instinct to survive as an individual, and several segments on the workings of human groupings.

Part III will touch upon the topics of human sexuality, morality, religion, economics, and politics. Throughout all of these discussions, particular attention will be paid to the ongoing conflict between self-serving and altruistic behavior. All of these topics will be explored with an evolutionary perspective in mind.

The final chapter in the book will apply a *paleo perspective* to list of what one may consider commonplace, everyday, and mundane wonderings. This last section represents an attempt to reset the reader's mindset, as he or she questions the world around him or herself.

Consider one final point. Throughout the discussions, *The Paleo Perspective* argues that many human behaviors are genetically based. This is an important concept and one that is critical to the thesis of the book. When contemplating if a particular behavior is innate, one should look for and consider the following:

- The behavior is more or less universal
- The behavior requires no special training
- The behavior is repeated over and over again
- The behavior starts in early childhood
- The behavior is found in the non-human animal world
- There is evidence of the behavior in man's ancient past
- Modern science has identified the particular genes responsible for the behavior in question

Please note that a behavior need not satisfy every one of these criteria to be considered genetically based.

PART I
THE BASICS

Our Paleolithic Design

CHAPTER 1

THE PROBLEM

"I can calculate the motion of heavenly bodies but
not the madness of people." - Isaac Newton
(English mathematician, astronomer, and physicist)

Sometimes it seems like the world has gone crazy. All too often we read about murders, mass killings, and wars. There are all sorts of crimes and acts of violence committed in small towns, big cities, and in every nation around the globe. There is racial conflict, ethnic conflict, religious conflict, tribal conflict, and every kind of social unrest. According to the *Substance Abuse and Mental Health Services Administration (2014)*, about one out of every five Americans suffers from a diagnosable mental disorder. Poverty, hunger, homelessness, income inequality, and pollution are all part of life in modern times. Is this sad state of affairs the way that it has always been since the very beginnings of our species, some 200,000 years ago? Are we incapable of getting along with each other? Are we incapable of functioning as a viable and fruitful species?

As crazy and dysfunctional as the world often seems, it should also be pointed out that alongside of all this negative and destructive behavior, human beings are also capable of exhibiting incredible acts of compassion, kindness, and altruism. People pull strangers out of burning cars and buildings. They donate blood, bone marrow, or a kidney to perfect strangers. Many people volunteer at soup kitchens, food banks, or make donations to various charities. It is noteworthy and significant that sometimes the very same people who are capable of these acts of altruism are also guilty of racial hatred, religious intolerance, or acts of violence.

What gives? How can we make sense of this sometimes frightening and confusing state of affairs in which we find ourselves? One might wonder if there are some fundamental and consequential flaws in our genetic design.

The problem is not that we are a poorly designed species—nature and natural selection are very good at what they do. The problem is that times have changed, but we have not. Genetically speaking, we are the same creatures that we were roughly 200,000 years ago (Carl Zimmer, *Evolution: The Triumph of an Idea*, 2006). Evolution proceeds very slowly, and there hasn't been enough time for us to change significantly in such a comparatively short period of time. (See important note at the end of the chapter.)

The fact is that we were simply not designed for modern times—we were designed for Paleolithic times and Paleolithic circumstances. What was genetically appropriate and adequate for Paleolithic times is often wanting and deficient for modern times. Our current dilemma is that there is a profound mismatch between our genetic design and "modern" circumstances. In many ways modern circumstance overwhelms us, leaving much of the world in disarray. To be clear, the term "modern" is being used to mean less than roughly 12,000 years ago, with the birth and proliferation of "civilization."

To make sense of and understand the mayhem and contradictions that we see all around us, we must begin to think of ourselves as an ancient species—a Paleolithic species, with Paleolithic bodies and Paleolithic minds—living in a whole new world. Modern circumstances are entirely new. Mankind has never been here before. While humans are in general a very adaptable species, modern civilization puts a huge strain on our genetic template.

It is true that many books have been written exploring the stresses that modern man has experienced over the last several centuries, *Future Shock* by Alvin Toffler (1970) comes to mind. But what we are exploring here is a much more fundamental sociological conundrum. *The Paleo Perspective* is comparing human life during Paleolithic times (from about 2.6 million years ago to about 12,000 years ago)—when we humans sheltered in caves and lived life primarily as communal hunter-gatherers—to what we now consider modern times (less than 12,000 years ago). We should note that humans and their ancestors lived in the communal hunter-gatherer format for about 84,000 generations, as compared to only about 350 generations in the modern format.

What have we inherited from our genetic past? How far have we drifted from our roots, our hunter-gatherer origins? What happens when we Paleolithic humans try to cope and survive in a drastically altered, modern world? What are the social, economic, and political ramifications of this mismatch? These are the central questions of this book.

Let us close this introduction by considering a phenomenon sometimes referred to as "zoochosis," or animal madness. There have been many articles and books written about what happens when animals are forced to live in unnatural habitats. A case in point was a polar bear named Gus who lived in the Central Park Zoo in New York City.

> *"In the mid-1990s, Gus, a polar bear in the Central Park Zoo, alarmed visitors by compulsively swimming figure eights in his pool, sometimes for 12 hours a day. He stalked children from his underwater window, prompting zoo staff to put up barriers to keep the frightened children away from his predatory gaze. Gus's neuroticism earned him the nickname "the bipolar bear," a dose of Prozac, and $25,000 worth of behavioral therapy."* (Laurel Braitman, *Animal Madness,* 2014)

In many ways modern man is a victim like that polar bear in the zoo. But we are victims of our own doing—our own "success." Modern circumstances are extraordinary circumstances, and in many significant ways, an unnatural setting for man. Human beings are no different from other animals. When forced to live in circumstances that they were not designed for, all animals will exhibit abnormal behaviors in response to these unnatural settings. This is the central thesis of *The Paleo Perspective*.

Note: It is important to point out that there are several examples of some very specific genetic variations that are present in today's population that did not, as far as we know, exist 200,000 years ago. For instance, there were no blue-eyed humans 200,000 years ago. That particular trait originated somewhere between 6,000 and 10,000 years ago. Lactose tolerance goes back about 4,000 years. And light skin goes back about 7,000 years. Changes such as these are usually caused by a variation in a single gene (HERC2 for blue eyes, and LCT for lactose tolerance). Since humans have roughly 20,000 genes, these genetic alterations represent an extremely small over-all change (.005%). In so much as any two humans differ genetically by about .1%, these kinds of changes represent an insignificant and inconsequential variation within the human population.

No rational person would argue that people with blue eyes, green eyes, or brown eyes are fundamentally different from one another, or that being lactose tolerant changes your human nature in some significant way. And unless you are a bonafide racist, no one believes that the color of your skin alters your humanity in some meaningful way.

These kinds of genetic changes do not represent revisions in basic human nature. There are genetic variations across any given population. Everyone is indeed different. But we are similar enough to all be considered human.

What we are talking about here is the lack of fundamental changes in humankind in the last 200,000 years. Any one of us today would go unnoticed in a Paleolithic cave. And if a Paleolithic man were raised in modern times, he would be no different than you or I. As previously stated, our species has existed in its present form for about 200,000 years.

CHAPTER 2

THE PALEO PERSPECTIVE

"Man is the only creature who refuses to be what he is."
- Albert Camus (French philosopher, author, and journalist)

According to the *Linnaean* scientific classification system, "man" is classified as follows:

Kingdom – Animalia
Phylum – Chordata
Class – Mammalia
Order – Primate
Family – Hominidae
Tribe – Hominini
Genus – Homo
Species - Homo sapiens

We are part of the animal kingdom, just like every other animal that inhabits this Earth—big or small, fish or fowl. Some might be offended by classifying man as an animal, arguing that he alone is special, and in some very real way, different, separate, and above the rest of Earth's inhabitants.

Others might argue that all of God's creatures are special, and therefore we shouldn't feel degraded, but instead feel honored to be included in such divine company.

Putting aside religious arguments for the time being (religion will be discussed in a later chapter), let us acknowledge the fact that although the scientific community may consider man to be special, it does not consider him to be separate from the rest of the animal kingdom. It does us no good to operate under any delusions we might have regarding the nature of man. We have to be willing to look at ourselves with the same level of objectivity that we look at other species. As far as we know, humans haven't significantly changed for the last 200,000 years. Failing to acknowledge man's genetic or intrinsic nature puts us at a disadvantage when facing life's challenges, and life is challenging enough.

Evolution as a Worldview

A significant number of the American public does not believe in Darwin's theory of evolution. According to a 2014 Gallup poll, 42% do not. These individuals believe that the Earth and all its inhabitants were created less than 10,000 years ago in their current form. Of the people who do believe in evolution, it would be safe to say that the vast majority of them do not walk around with thoughts of Darwin's theory swirling around in their head. The theory of evolution is not part of their everyday mindset. They go about their daily lives having relegated Darwin's theory to a remote corner of their memory, like a science book gathering dust on a library bookshelf.

But the theory of evolution is arguably the most profound and relevant scientific discovery to ever come along. It is the basis for any scientific worldview, and it is ever-present in my mindset. And so every topic in every chapter of this book deals with this perspective—what I am calling a *paleo perspective.*

"Our animal origins are constantly lurking behind, even if they are filtered through complicated social evolution." - Richard Dawkins (English ethologist, evolutionary biologist, and author)

The term "paleo" refers to something old or ancient. For example, the term Paleolithic refers to the earliest or oldest Stone Age: from about 2.6 million years to about 12,000 years ago. The term Neolithic would suggest the most recent Stone Age: from about 12,000 to 7, 000 years ago. It turns out that "modern" man is not really that modern, and in fact exhibits many ancient proclivities and urges. For the most part, modern man is more or less the same creature that he was 200,000 years ago. And so modern man is still a Paleolithic man.

Human evolution has been going on for about the last 6 or 7 million years, when man split from his last common, ape-like ancestor. That's a very long time. A mere two hundred thousand years is too short a time span to expect much in the way of evolutionary change. Human nature is more or less the same now as it was back then. Once you accept and are mindful of our evolutionary past and our evolutionary inheritance, you tend to look at the world differently—it changes your perspective.

To examine this all-important mismatch between our Paleolithic design and modern times, we must first take a brief look at what life was like 200,000 years ago.

Man: the Hunter-Gatherer

Except for about the last 12,000 years, hunting and gathering was presumably the lifestyle of our ancestors. Hunter-gatherer groupings sustained themselves by foraging, hunting, and fishing. Hunting involved both small and large prey, and probably included some scavenging when the opportunity presented itself. Weaponry initially included stone axes, with the use of spears, arrows, and some finer cutting tools coming later on. Gathering (foraging) involved the search for edible plants, fruits, berries, nuts, and tubers. Most hunter-gatherer societies were nomadic, or semi-nomadic. Movement was necessary to prevent the over-utilization of resources in any particular area. These mobile groupings probably made use of natural rock shelters, or some type of temporary makeshift shelter.

Evidence suggests that these ancient hunter-gatherer groupings consisted of perhaps 20 to 60 (maybe as high as 100) individuals (*Science and Technology in World History*, McClellan, 2006). These small numbers suggest that everybody knew one another, and most likely consisted of family and extended family members. Collaborative hunting, foraging, and child rearing would have been absolutely necessary in such primitive and difficult times. Familiarity, trust, and tribal loyalties would be significant. (Please note that the use of the term "tribe" or "tribal" throughout this book is somewhat colloquial, and refers more to a social division in society and not the taxonomic category that is used in the Linnaean classification system.)

In such an intimate setting, cooperation and a strong sense of community would have been the order of the day. This strong sense of "us" was essential to ensure the evolutionary fitness of the grouping.

Besides trust, there were many other human emotions and sensitivities that were essential for the effective functioning of these primitive groupings. Emotions and feelings such as compassion, empathy, envy, embarrassment, guilt, love, pity, remorse, shame, and sympathy were all vital. They are all instinctual and genetically based, and are widely accepted as being part of human nature. Nature has seen to it that we are all born with a capacity for these sentiments. They are absolutely critical and indispensable for living in complex social settings. It is central to the theme of this book, and thus critically important to point out, that these sentimentalities are most effective in modulating and shaping human behavior in small, intimate settings—like the ones our ancestors lived in.

Hunter-gatherer groupings tend to have an egalitarian social ethos. Because of their mobility, the acquisition of material goods would be at a minimum, with a rather equal distribution of possessions (wealth). No one member could be significantly "richer" than other members of the troupe. Karl Marx defined this socio-economic system as "primitive communism" (Scott, John, Marshall, and Gordon, *A Dictionary of Sociology*, 2007).

In some ways this ancient setting can seem rather idyllic. Along those lines, there was an article that appeared in *The New York Times* ("The Great Affluence Fallacy," August 9, 2016) in which author David Brooks references Sebastian Junger's book, *Tribe*. Brooks makes reference to the reluctance of European settlers, when captured by Native American Indians, to return to their European villages. They much preferred the hunter-gather life-style of their captors. In contrast, most captured Native Americans returned to their hunter-gatherer villages when given the chance.

"Human beings will be happier— not when they cure cancer or get to Mars
or eliminate racial prejudice or flush Lake Erie but when they find ways
to inhabit primitive communities again. That's my utopia."
- Kurt Vonnegut (American author and satirist)

Today there are almost no true hunter-gatherer societies left on Earth, although the Bushmen of the Kalahari Desert come close. Like it or not, this ancient hunter-gatherer setting is the social milieu that our DNA was optimally designed for. Today's large cities and states do not provide the same level of intimacy and community that those archaic hunter-gatherer groupings provided. Ironically, one of the most significant ramifications of ever-larger modern municipalities is the loss of a sense of community. In many ways, human nature was simply not designed for modern times.

A Paleo Perspective

The Paleo Perspective is not just about looking at the problems caused by the discrepancy between Paleolithic times and modern times. It also endeavors to peel away modernity's many layers and assumptions, revealing human nature's ancient identity. A *paleo perspective* is a scientific and anthropological perspective, with human evolution as its focal point. It is about using an anthropological lens to look at human nature, and to examine who and what we are as a species.

The fact is that our current human design is not new, and is the result of millions of years of evolution. There are even some parts of our genome that are hundreds of million years old. For example, we share some DNA with the shark (154 genes with the elephant shark). Sharks and humans share certain physiological processes, like the way we have sex and how we protect our bodies from disease (Jennifer Vargas on *ABC Science*, May 30, 2007).

Perhaps the best way to introduce the reader to the concept of a *paleo perspective* is to give several examples of how to apply it to some recurrent and ubiquitous social problems and phenomena. These examples will give you a sampling of the *paleo perspective* in action. As you read each example, please note how our ancient but contemporaneous instincts are central to each problem. Subsequent chapters will discuss many more important and relevant issues in greater depth.

Racism and Hate Crimes – Life for a Paleolithic grouping was basically "us" against the world. Besides the stresses of nature, ancient human bands would often find themselves in competition with other human bands—and perhaps even human sub-species (like Neanderthal)—for food resources and territory. Having a strong affinity for your own tribal kind, and at the same time, being distrustful and fearful of those who are in some way—*any* way—different from you, helps your group cooperate and collaborate more effectively in the defense of your desired resources and territory. The group that sticks together better has a better chance of survival against nature and against competition.

And so, even in modern times, it is quite natural for us to cling to that which is familiar ("us") and be suspicious and apprehensive of that which is different ("them"). Racial, religious, ethnic, and cultural differences are all instinctual hurdles for all of us to overcome. Modern life, with its multicultural cities and countries, presents a problem to our Paleolithic mind. And so acknowledging our inherited prejudices is a first step in trying to deal with this pervasive and universal problem.

Teen pregnancy - Other than the obvious fact that engaging in sexual activities can be extremely pleasurable, why is teen pregnancy such a persistent issue in modern society? If we look back to "paleo–times," we discover that the average life span was about 30 years (*Encyclopedia of Population*, 2003). It stands to reason then that sexual intercourse and childbearing would have to start soon after the onset of puberty for both males and females. Teen fathers and teen mothers would not have been problematic. In fact, having children much later in life would be more problematic, given the limited life spans. Having a child at 25 would severely limit the amount of time one had to fully raise a child, and to instill in that child the necessary life skills the child would need to survive independent of its parents, and start a family of its own. In Paleolithic times, teen pregnancy was probably the order of the day.

However, teen pregnancy can be problematic in modern times. Today's youngsters require a prolonged period of training and education, and

12

aren't usually prepared for adulthood until they have finished high school or perhaps college. Nowadays, youngsters aren't fully independent until they are in their twenties. Considering that modern life expectancies for most countries is roughly 70 years (United Nations' "World Population Prospects," 2012), this leaves plenty of time to raise children, even if one gives birth in their late twenties or even late thirties.

So the problem is how to keep post-pubescent youngsters from doing what comes quite naturally to them, long enough for them to get the proper education and training necessary for modern adulthood. This is a clear case of our Paleolithic bodies and Paleolithic libidos being out of sync with the realities of modern life.

Obesity - During Paleolithic times securing enough food was an ongoing struggle. There was no capacity to store meats or edible vegetable matter. Being nomadic meant that you couldn't carry large quantities of food with you as you traveled, even if you did have a way to preserve it. If your clan had a successful hunt, you probably ate what you could, knowing that the meat wasn't going to keep for very long. Ancient humans needed animal fats to provide them with lots of calories for energy and to help keep them warm. And if you came across a good patch of berries, you most likely gorged yourself on berries.

But various foods were not available every day and in every season like they are today. There was very little chance of over-consuming fats, sweets, or carbohydrates on an ongoing basis because they weren't available everyday, every week, or every month of the year.

In addition to the sporadic food supply, there was the fact that you and your clan were nomadic, constantly on the move. You couldn't possibly maintain this hunter-gatherer, nomadic life-style and be seriously overweight. It is unlikely that there were many obese cavemen.

Nowadays, most people can walk into a supermarket and get most anything and everything they want, no matter the day or the season. Modern man has the opportunity to overindulge in meats, sweets, and starchy foods. Add to this the fact that most of us lead very sedentary lifestyles, and it is easy to understand how obesity and type II diabetes can be so problematic in modern times, especially here in the USA.

Having cravings for sweets, carbohydrates, and fats served primitive man well because they enticed him to take advantage of opportunities when they presented themselves. The unpredictable nature of Paleolithic man's food sources did not allow for the ongoing misuse of these resources. On the other hand, most of us now have continual access to a wide variety of foods, allowing us to overindulge our Paleolithic urges.

Gang Formation - To a very large degree, Homo sapient success as a species lies in his proclivity to form complex, highly cooperative groupings. These Paleolithic groupings (20 – 60 individuals, often related) hunted together, foraged together, slept together, raised their offspring together, fought together, and survived together. Man instinctively understands that his survival is greatly enhanced when he is part of a group. This predilection for group formation, cooperation, and collaboration is encoded in our Paleolithic DNA. Remember, our present DNA design is the original Paleolithic model. There has not been enough time for a new, updated design to replace the old one.

And so we still have that urge to form a group, join a group—be part of a group. We look for shared values, interests, mindsets, and worldviews. Gangs in particular draw people with shared mindsets and worldviews. Many people join gangs to commiserate with those with the same racial or ethnic background. They also join gangs for the protection that a group offers them, especially in crime-ridden areas.

The Paleo Perspective also helps us understand better why many individuals want to enlist in the army, navy, or marines, or be part of some sort of militia or jihadist group. They do so to be part of something larger than themselves, to fight for a group cause. Just like Paleolithic man, who sometimes did altruistic things for the group's benefit, people today often consider their group's survival or their group's cause to be more important than their own survival.

Joining a gang or a terrorist militia is a misdirected response to something that we all feel from time to time—a Paleolithic urge to be part of a group. We feel satisfaction in the intimacy, the camaraderie, the sense of security, and the perception that we are part of something larger and more important than ourselves.

Understanding why people participate in some awfully bad groups is not to condone some really bad behavior. Everybody has a brain, a conscience, and free will. We should note that the understanding and acceptance of innate human proclivities is crucial to the true exercise of free will.

Soap Operas and Reality TV Shows - Why are soap operas, novels, and gossip so popular with many of us? Even people who don't read novels or watch reality TV shows find it hard to look the other way when there is a "social train wreck" unfolding in full view.

Man is a highly socially complex animal—by far the most socially complex animal on the planet. Many types of ants lead socially complex lives, but they rely on pheromones (chemicals) to direct their behaviors. We depend on our brain to interpret and negotiate complicated social situations. Our lives are filled with emotions such as love, hate, sympathy, and remorse. People lie, cheat, and steal. They plan, they scheme, and they fake. All of this requires a large degree of "social intelligence," and at the very least it requires that they be aware of and tuned into all of these social machinations. In fact, the group's operation and success depends on all the players being able to comprehend and manage all of these social phenomena.

And so our ancestors needed to be tuned into all of the social nuances involved with group living. Paleolithic man's DNA needed to provide him with an innate predilection to be interested in, and focus in on, anything and everything social—especially problematic social happenings. Paying attention to negative news is advantageous in that it helps us focus on any problems that need our attention and need to be solved. It is a Paleolithic urge and a social necessity to be "nosey."

It is interesting to note that in the very popular TV series, *Survivor* (CBS, 2000 to 2016), contestants are divided into small tribes. These tribes are then pitted against one another in primitive survival scenarios. The popularity of this series is testament to the fact that man is instinctively drawn to human-drama situations. It is also particularly interesting that this series is reminiscent of our hunter-gatherer past.

Love of Music and Dance - The Paleolithic usefulness of music and dance will be discussed in more detail in a later chapter, but let us start off by saying that they both are instinctual. They are part of our genetic makeup. Music and dance are practiced in every culture, in every corner of the world. This is true now, it was true 5,000 years ago, and it was true 200,000 years ago for our Paleolithic ancestors. One does not need formal training to appreciate music or to move to the rhythms of the beat. Humans instinctively tap their fingers and feet to the beat. They teach themselves how to play the guitar and the piano. Music and dance are uniquely human.

Is music a gift from God? Is it an example of evolutionary serendipity? Perhaps we just lucked out. What then is the Paleolithic purpose to man's affinity to and love of music and dance?

Simply put, music and dance helps keep human groups together. When we sing together, when we dance together, we get a feeling of camaraderie, warmth, conviviality, and goodwill. These shared experiences are the things that help a group feel like a cohesive whole. These are the things that bind. It is true today and it was especially true in Paleolithic times when group cohesion was absolutely essential for survival. Music is no serendipitous accident. It is an evolutionarily useful phenomenon.

Homosexuality - I am using homosexuality to illustrate the old adage that sometimes the exception proves the rule. Science is yet to agree on an ironclad explanation for human homosexuality—or homosexuality in any other species. There are some theories about the usefulness of homosexual behaviors in bonobo chimps that have something to do with the reduction of aggression. (Brian Hare and Vanessa Woods of Duke University, 2014) Obviously, human homosexuality does nothing to ensure the continuation of our species through the process of sexual reproduction. But the percentage of homosexuals is not large enough to jeopardize a global population growth rate of about 1% - 2% per year (worldometer. info). After all, we are at about 7 billion and counting!

A *paleo perspective* on homosexuality is that evolution (nature) can in fact be adventitious at times. Perhaps human homosexuality is an illustration of the fact that our evolutionary design need not be perfect. It just needs to be good enough to ensure our survival. This is a good example of the fact that everything cannot be explained by looking for evolutionary purposefulness. Nature is not perfect—stuff happens.

The idea that our human design is far from perfect was alluded to in an article that appeared in *The Wall Street Journal* on April 14, 2018. (Dr. Lents, "The Botch of the Human Body") Dr. Lents—a professor of biology at John Jay College in New York—writes:

> ***"Over the past decade,*** *geneticists and biologists have learned more about our evolution than we ever thought possible. Not all of it is pretty. For example, the DNA in our cells is littered with huge stretches of repetitive, useless gobbledygook. Our chromosomes also harbor thousands of viral 'carcasses,' baggage left by infections that our ancestors fought millions of years ago.*

> *"Evolution has not perfected our species—far from it. The human body, wondrous and beautiful as it may be, is cluttered with glitches and inefficiencies, the messy byproducts of evolution's creative process. Natural selection is a blind, groping process, one that frequently produces terrible problems in addition to workable prototypes.*

> *"Our many design glitches, from weak knees to poor sinus drainage to infertility, highlight the randomness of evolution."*

And so as we try to apply the principles espoused in *The Paleo Perspective,* let us not think of evolution or nature as some sort of perfectionistic "watchmaker."

"One of the basic rules of the universe is that nothing is perfect. Perfection simply doesn't exist. Without imperfection, neither you nor I would exist."
    - Stephen Hawking (Physicist, cosmologist, and author)

CHAPTER 3

IN THE BEGINNING

The Big Bang

The Paleo Perspective is a scientific perspective, so let us start from what science considers to be "the beginning." According to generally accepted scientific theory, the universe is about 13.7 billion years old. In fact, science has no evidence of anything—matter, energy, space, or time—existing prior to 13.7 billion years ago. So, approximately fourteen billion years ago, something we now call the "Big Bang" happened, and in an instant all the things that we are familiar with, matter, energy, space and time, came into existence.

What existed before the Big Bang? Nobody really knows. It is very difficult for human beings to conceptualize what could have existed before space and time came into being. Some people who self identify as "religious" or "believers" might say that God existed before the Big Bang, and that God is responsible for the Big Bang and everything else that followed. Agnostics and atheists alike might admit that trying to conceptualize the non-existence of space and time without the concept of a "creator god" is vexing to say the least.

Maybe there is no such thing as a "beginning" or an "end." Maybe those things are constructs of the human mind, having evolved in a Newtonian world (under classical, everyday laws of motion and mechanics). Perhaps we can think of ourselves as having a neurological "operating system" hardwired to accommodate only Newtonian mechanics. But reality turns out to be more complicated when we introduce some of the concepts of relativity and quantum mechanics. Relativity and quantum theory blows most people's minds. It is tough to think "outside the box" when you have evolved to think and function "inside the box." However, some are better at it than others. Perhaps people like Albert Einstein and Stephen Hawking were better at thinking outside the box than most people—even most other scientists. But let's get back to "the beginning."

After the Big Bang

What do scientists think happened right after the Big Bang? For about the first 1/100,000,000,000,000,000,000,000,000,000,000,000,000 of a second, the universe expands by undergoing a spectacular unfolding known as "inflation." This is also very tough to conceptualize—"space" itself expanding. Most people immediately start to wonder what was beyond space itself? More space? No one knows for sure, but it would be illogical to declare that there is more "space" beyond space itself. It is the kind of concept that is difficult to wrap your mind around, but let's get back to our timeline.

After one whole second, protons and neutrons are formed. After about three minutes the nuclei of the first elements—hydrogen and helium—are formed. It takes 380,000 years for whole atoms to form and 30 million years for the first stars to form. After 200 million years our own Milky Way Galaxy forms, and finally after about 9 billion years the Earth itself is formed. That almost brings us up to the present.

Instead of looking forward from the Big Bang (as we have been doing) let's now reverse things and look back from the present. Let us look at how long ago various important developments occurred.

About 3 billion years ago some basic cell organisms developed and about 1 billion years ago multicellular organisms developed. The first fish appeared about 500 million years ago and primitive insects appeared about 350 million years ago. Mammals appeared about 200 million years ago and about 50 million years ago primitive monkeys appeared. About 25 million years ago apes split from monkeys and somewhere between 10 and 6 million years ago man's ancestors split from the ape.

Man's ancient ancestors are sometimes referred to as "hominins." Creatures such as Australopithecus, Homo erectus, and Homo

Heidelbergensis would be examples of hominins. There is evidence of stone tool fabrication going back approximately 2.6 million years and the use of fire going back about 500,000 years. Anatomically modern man (Homo sapiens) goes back about 200,000 years. So for all practical purposes, human beings—in our present form—have been around for about 200,000 years. If one of these "cavemen" were to pass you by on the sidewalk—dressed appropriately of course—you wouldn't give him a second glance. The more technically correct term for our caveman is "Paleolithic man." Let us continue our journey and get to the present as quickly as we can.

Civilization

Paleolithic man was basically living the life of what we would call a "hunter-gatherer." He was basically nomadic or semi-nomadic, hunting and gathering fruits, berries, and edible plants. Things really didn't change all that much for most of the 200,000 years. There are estimates that perhaps around 60,000 years ago Homo sapiens (man) began to migrate out of Africa. The cave paintings in Altamira Cave in Spain go back about 20,000 years. About 10,000 years ago man starts to domesticate plants and animals and starts to establish settlements, villages and towns. This is generally referred to as the "Neolithic" or the "Agricultural" Revolution.

Great ancient civilizations begin to develop. For example, Sumaria establishes itself in the Middle East in what is today Iraq. Perhaps this was the "Garden of Eden" of the Old Testament (Hebrew bible). Evidence of the use of the wheel goes back about 6,000 years and ancient writing systems go back around 5,000 years. Civilization is now well established, and man is not only the master of his fate, but has dominion over the entire earth and all its inhabitants. And so here we are, in the relatively recent present.

Big History

What we have just described is sometimes referred to as a kind of "Big History." It is a scientific account of things from the very beginning (Big

Bang) to the present. We have obviously left out the vast majority of facts, but have given you a very small sample of something called the *Big History Project* (created by David Christian and Bill Gates in 2011).

We start with *Big History* because man has always sought to find his place in the cosmos. History is littered with "big stories," stories about how the earth got here and how man came to be. The "creation story" we have just described is based on widely accepted scientific data. There are, however, various religious creation stories that some people take quite literally. For instance, some believe the earth to be only about 6,000 years old and that God created man in his present form instead of having evolved over millions of years.

The scientific account is a more complicated and intricate story than Genesis (the first book of the Hebrew bible), but it too attempts to answer the really big questions. This scientific story is more of a "how" story than a "why" story. And science is constantly tweaking some of the "facts"—that is the nature of science. We shouldn't lose faith in science because "new facts" replace some of the "old facts." Science is a search for knowledge, and the search goes on. It is a process and one should have faith in the process.

When we contemplate the details of this fact filled, scientific story, we get a sense of our place in the great cosmos. It can perhaps make us feel small, but it need not make us feel unimportant. Ancient man felt he was the center of the universe, but modern science is not sure there is even a center to the universe. That being the case, we can think of the Earth and mankind as being on a par with everything and everywhere else in the universe. As part of the cosmic fabric, man is not separate, but is connected by the laws of physics to every other part of the cosmos.

Timeline

For those of you who feel more comfortable with a more concise chronology:

- 0 – The Big Bang. Space and time are created.
- 1/100,000,000,000,000,000,000,000,000,000,000,000,000 of a second - The universe expands, undergoing a spectacular acceleration known as inflation.

- 1 second – The first composite particles, protons and neutrons form.
- 3 minutes – The nuclei of the first elements, hydrogen and helium, form.
- 30 million years – Stars first appear in the universe.
- 200 million – Our galaxy, the Milky Way, forms.
- 9 billion years – The Earth forms.
- 10 billion years – Life begins on Earth.
- 13.5 billion years – Early humans evolve in Africa.
- 13.7 billion years - The "present!"

"As we shall see, the concept of time has no meaning before the beginning of the universe. This was first pointed out by St. Augustine. When asked: 'What did God do before he created the universe?' Augustine didn't reply: "He was preparing Hell for people who asked such questions."
- Stephen Hawking

CHAPTER 4

EVOLUTION

*"Today the theory of evolution is about as open to doubt
as the theory that the Earth goes round the Sun."*
– Richard Dawkins

The Paleo Perspective holds that Darwin's theory of evolution is a scientific fact. And as stated in the introduction, this perspective involves looking at the human condition through the lens of evolution. Let us therefore start with a brief synopsis of the theory. Keep in mind that this book is not intended to be a textbook on evolution or anthropology. However, to employ a *paleo perspective*, it is essential that we have at least a rudimentary understanding of evolutionary theory.

Anybody can put their hands on the information presented in this chapter by consulting just about any book on evolution. Several examples are *Evolution: The Triumph of an Idea* by Carl Zimmer (2001), *Human Evolution* by Roger Lewin (2005), and *The Social Conquest of Earth by* Edward Wilson (2012). Also, anyone can sit down with their computer and "Google" hundreds of sites on the topic. Several examples are *Academia.edu, American Scientist.org, Humanorigins.si.edu, Columbia.edu*, and *Britannica. com*. The point is that information on evolution is easily accessible and widely accepted by the vast majority of scientists—both here in the USA and around the world.

A Relevant Theory

*"Some people would claim that things like love, joy and
beauty belong to a different category from science and can't
be described in scientific terms, but I think they can now be
explained by the theory of evolution." - Stephen Hawking*

The theory of evolution is consequential, profound, and particularly relevant. Its relevance seems obvious because it has important implications about us, Homo sapiens. But the theory of evolution also explains how all other living things came to be, from simple one-celled creatures to the vast array of diverse and complex creatures—past and present.

It explains how each life form is a result of a modification of some earlier life form through the process of genetic alteration and natural selection. Each species, past and present, was not designed from scratch—from the ground up. Each life form is a remodeling of some earlier life form. This is in sharp contrast to the idea that all current life forms were created in their present form (as described in Genesis).

Science tells us that each species has evidence of its shared evolutionary history within its DNA profile. Scientists (geneticists in particular) can see the evidence of this shared history—this shared ancestry—by analyzing and comparing the DNA of various species. For example, humans and chimpanzees have about 98% of their DNA in common. We share a common ancestor with chimps—an ancestor that lived about 7 million years ago. Humans share about 90% of their DNA with cats and 50% with bananas. These genetic similarities are evidence of our shared ancestry. Life has been constantly (albeit relatively slowly) evolving since the beginning, and continues to do so.

The principles and consequences of evolution are all around us such as drug- resistant bacteria, genetically altered foods, and the constantly-evolving winter flu virus. We can also see the evidence of our own evolutionary journey when we take note of some interesting characteristics of human nature.

For instance, studies have been done regarding what people look for when they are asked to select a leader. American presidents have tended to be about two inches taller than the average American male. Looking at the relative heights of American presidents from 1900 to 2011, we find that 18 were taller than their opponents, and 8 were shorter. It would seem that Americans generally prefer taller presidents.

How about voices? According to an article published in the *PLOS ONE* journal (Kofstad, Anderson, and Norwicki, "Perceptions of Competence, Strength, and Age Influence Voters to Select Leaders with Lower-Pitched Voices," August 7, 2015), "Voters prefer leaders with lower-pitched voices because they are perceived as stronger, having greater physical prowess, more competent, and having greater integrity."

Both of these examples suggest that we are expressing some evolutionarily based proclivities that were probably beneficial in Paleolithic times—the selection of bigger and stronger prehistoric group leaders.

A Profound Theory

"Evolution isn't just a story about where we came from. It's an epic at the center of life itself. Far from robbing our lives of meaning, it instills an appreciation for the beautiful, enduring, and ultimately triumphant fabric of life that covers our planet. Understanding that doesn't demean human life—it enhances it."
- Kenneth R. Miller (American author and molecular biologist)

Many scientific theories are critically important such as Newton's laws of motion, Einstein's theory of relativity, and quantum mechanics. But the very concept of evolution is profound because it helps explain how man came to be, and why the world around him is as it is. It involves who and what we think we are in relation to all the other living things that inhabit the Earth. It gives us a perspective on our place in the universe and helps us understand what it truly means to be human. As such, the concept of evolution has ramifications and repercussions regarding such issues as morality, religion, economics, and politics. (We shall discuss these issues later in the book.)

Evolution 101 – "Abbreviated"

We should make note of the fact that the theory of evolution is not a particularly new theory. Darwin published his *Origin of Species* in 1859. In spite of the fact that it is over 150 years old, evolution remains somewhat controversial within some religious circles—particularly ones that take various biblical creation stories as literal fact. Be that as it may, let us take a brief look at how evolution works.

What are the basic mechanisms and biological processes that make evolution possible? There are several things that must be true for evolution to work.

First, any living organism must be able to reproduce—to replicate itself. If life couldn't replicate itself, planet Earth wouldn't be teeming with the immense number and variety of living creatures that it does.

Secondly, life-form reproduction cannot be a perfect replication process. That is to say that every one of their offspring cannot be perfectly genetically identical to their parents. Offspring production must result in some genetic diversity within the specific population. Genetic diversity helps ensure the survival of a species because it facilitates evolutionary adaptation to changing environmental conditions. If cell reproduction were a perfect replication process 100% of the time, life as we know it would not exist. We will see a little later on in this chapter just how DNA replication and mutation help facilitate genetic diversity.

Lastly, since its creation, environmental conditions on Earth have been constantly changing—and will continue to change. Mountains rise and fall. Sea levels rise and fall. Rivers, lakes, and glaciers come and go. Rain forests become deserts. Tundra becomes forest. These geological, meteorological, and environmental changes are the driving forces behind evolutionary change. In order for the Earth's various life forms to survive in ever changing conditions, they must adapt, they must change—they must evolve.

Replication, Reproduction, and Mutation

"The capacity to blunder is the real marvel of DNA. Without this special attribute, we would still be anaerobic bacteria and there would be no music."
- Lewis Thomas (American physician, etymologist, and essayist)

As we have just noted, all life forms must be able to replicate themselves—to reproduce. Inorganic rocks form and break down into smaller rocks. Crystals increase in size—they grow. But these changes are very different from living cells dividing or plants growing. When a cell divides it creates a replica of itself. One cell becomes two cells, two cells become four, etc. This is possible because large, complex organic molecules such as ribonucleic acid (RNA) and deoxyribonucleic acid (DNA) are able to replicate themselves.

Science has made great strides in its effort to understand and replicate nature. Researchers have been trying to create a self-replicating molecule in the laboratory for some time. In fact, researchers have successfully synthesized the basic ingredients of RNA, a molecule from which the simplest self-replicating organic materials are made. Scientists at the Scripps Research Institute created molecules that self-replicate, evolve, compete, and breed (Robert Britt, "Life as We Know it Created in Lab," *Lifescience. com*, Jan 11, 2009). Most importantly, every once in a while one of these molecules makes a mistake in the replication process. Lo and behold, that's exactly what natural life forms do—they make mistakes.

These mistakes are referred to as mutations. A mutation is a change in DNA, the hereditary material of life. An organism's DNA affects how it looks and how it behaves. So a change in an organism's DNA can cause changes in all aspects of its life. A mutation can occur during the cell's normal replication process, but radiation and chemicals can also cause mutations. Let us look at how mutations and genetic variations within a species' population facilitate evolution.

Back in the middle of the nineteenth century, a monk named Gregor Mendel discovered the basic principles of genetics—how various characteristics are passed on from one generation to the next through DNA combination and replication. When a mutation occurs, the offspring will be in some way different than the parents. Small genetic differences between the parents and the offspring can accumulate in successive generations, so that after many generations, the descendants may differ significantly from their ancestors. This phenomenon is at the very heart of evolution.

Beneficial Mutations

If a mutation were to enhance the survival chances (sometimes referred to as "evolutionary fitness") of the offspring, that genetic variation would increase in frequency in the gene pool of that species. This genetic modification would continue to grow in frequency and become more common until the vast majority of the population inherited and exhibited the improved characteristic. This process is referred to as "natural selection."

An example of this kind of beneficial genetic variation—this evolution by natural selection—can be illustrated by the giraffe. Imagine a situation where the food supply for the giraffe, perhaps because of some climate change, was more abundant in taller trees. A giraffe born with a slightly longer neck (genetic variation) would have access to more food, increasing its chances of survival, and pass on its "long neck" genetic variation to its many offspring. At the same time, giraffes with shorter necks would have access to less food, be less robust, and as a result would have fewer offspring. It is not too difficult to imagine that given enough time, and given the obvious advantage that the taller giraffes would have, the giraffe's neck would evolve to its present, rather long state. It should also be pointed out that if the climate didn't change, and the food supply had remained abundant low to the ground, there would be no advantage in having a long neck. In fact, it might be a disadvantage (awkwardness) and that "long neck" genetic variation would not have become more common but instead would have died out.

Another case to consider is that of the chameleon. At some point in time a mutation occurred that allowed the chameleon to alter its color to match its surroundings. This genetic variation most likely happened in small steps and over great intervals of time. It is easy to see how the chameleon's ability to blend into its surroundings would make it more difficult for predators to see it, and thereby enhance its survivability—its evolutionary fitness. On the other hand, if changing color made it easier for predators to see chameleons—and eat them—the genetic variation (mutation) that allowed them to change color would have quickly died out.

Yet another example of evolution by natural selection is illustrated by the honeybee. A recent study done by Geraldine Wright, a honeybee brain specialist at Newcastle University in England, demonstrated that the naturally caffeine-laced nectar of some plants enhances the learning process for bees (James Gorman, "Nature That Gives Bees a Buzz Lures Them Back for More," *The New York Times*, March 7, 2013). These bees are then more likely to return to those caffeinated flowers. The result is that more of these flowers get pollinated and more bees get fed. This benefits both the flowers and the bees. This is natural selection at work.

Here is yet another fairly recent illustration of evolution through natural selection. Cockroaches are considered to be outstanding survivors, with enough evolutionary tricks to have survived and thrived for the last 350 million years. According to an article that appeared in the May 24, 2013 issue of *The New York Times* (James Gorman, "Wily Cockroaches Find Another Survival Trick: Laying Off Sweets"), researchers in North Carolina discovered that some populations of cockroaches actually managed to switch their internal chemistry around.

Pesticides are normally laced with glucose (a type of sugar) to entice cockroaches to devour the poison. But these cockroaches switched their internal chemistries around so that the pesticides tasted bitter to them instead of sweet. This enhanced their survivability because they were no longer attracted to the sweet taste of deadly pesticides. This is a good example of evolution in response to an environmental change (albeit a human-engineered environmental change)—the availability of sweet tasting, deadly pesticides.

Consider one more illustration of evolution at work. Imagine a large piece of land that was covered with various trees, bushes, plants, and grasses. If the climate suddenly became drier, the plants that required lots of water would suffer and eventually their numbers would decline—or maybe disappear altogether from the setting. In contrast, the plants that liked to thrive in a drier environment would increase in number, and in time this land area would look quite different—more of some plants, less of others.

But if this climate change happened very slowly, it would give some of the plants enough time to change—to evolve into plants that could survive with less water. This would happen through the process of mutation. This mutation process would not happen in one gigantic step—like a water loving plant giving birth to a cactus. More likely, plant seeds that were slightly better adapted to a dryer setting would survive better and pass their genetic variation onto their offspring. After much time and many generations, these plants might look quite different and be capable of thriving in dryer climates. However, if the climate change happened too quickly, the process of genetic variation through mutation would not have enough time to accomplish the necessary modifications and extinction might be the result.

Although mutation rates vary from species to species, given enough time these genetic mistakes account for the ever-changing and diverse forms of life on earth—past and present. Evolution is as natural as any other biological process. It continues today all around us—unabated and unstoppable!

Because evolution is an ongoing process it is tempting to imagine that it has a direction, or a preordained purpose. It is tempting (and somewhat egocentric) to consider man to be the endpoint or goal of evolution. Humans do tend to think of all other species as being more primitive than they are. And so the layman sometimes wonders why didn't other species (like the chimpanzee) continue to evolve further toward the "finish line" (Homo sapiens).

Every species has had to make modifications in its design to meet the environmental challenges in their particular eco-niche. But a shark has no need to be more human-like—it is well adapted for the sea. An eagle has no need to be more human-like—it is well adapted for the sky. Conversely, there is no need for humans to grow gills or wings.

Today's present-day chimpanzee is different from its ancient ancestors. Chimps have been evolving to their present form just as humans have been evolving to their present form. Each species' current design is appropriate for its current eco-niche. Evolution's only direction is toward survival, not toward a particular predestined design. In the case of evolutionary designs—"one size does not fit all."

The Evolutionary Maze

It might be helpful to think of a maze when we are trying to gain insight into the evolutionary process. Generally speaking, at any particular location on a maze, one has several options, or directions, in which to move. Picture if you will each species being at a different location in this figurative maze. Each location in the maze represents a particular eco-niche and how a particular species has adjusted and evolved—physically and behaviorally—to survive in that niche.

Take for example the African elephant. The elephant has developed an extremely long and versatile nose (trunk) to help him gather food and suck up water. He has huge ears to help him keep cool and thick skin and large tusks to help protect him from big cat predators. He has developed all of these features to help him survive in the particular environment he finds himself. All of these adaptations, along with the particular set of environmental conditions that the elephant finds himself, represent a particular location on our figurative maze.

If environmental conditions change, the elephant may change, but his options are limited. He might get a little smaller if food becomes more scarce. He might grow smaller tusks to make him less appealing to ivory hunters. His skin might get a bit thinner if it gets hotter and there are fewer predators around.

But his location in the maze means that the elephant is not about to swap his trunk for a set of gills. Developing gills is an option for a species on a completely different part of the maze: one where there is a sizable body of water around, and one where the species in question has already made some evolutionary adaptations that lend themselves to life in the water— webbed feet, for instance. Similarly, there is no chance that a shark will develop a long neck to feed on tree top leaves or fruit like the giraffe did.

Each species has moved through that evolutionary maze, changing its physical and/or cognitive attributes in response to changing environmental circumstances. And like being in a maze, each species only has a limited number of options (directions and turns) to choose from to make the necessary design modifications to enhance survivability.

Man's Position on The Evolutionary Maze

And what of man's position on that evolutionary maze? Man is not a terribly fierce animal, physically speaking. He is not the biggest, strongest, or the fastest animal on the planet. He doesn't possess sharp fangs or long claws and doesn't possess special physical features like wings, protective needles, or poisonous flesh.

A long time ago, our ancestors took a turn in that evolutionary maze. It wasn't a turn toward greater physical prowess, but a turn towards greater intelligence. Given the enormous number of different species, and the large number of possible evolutionary adaptations, one might not be too surprised that eventually one species might navigate the evolutionary maze by making cognitive enhancement its signature move. Perhaps it was just a question of time.

The Good, the Bad, and the Funny

As stated earlier, evolution is a relevant and profound theory. But it is also quite controversial in America, even to this day. There are many interesting quotes on evolution from famous people—scientists, comedians, and politicians. Some speak highly of the theory, some are ardently critical of the theory, and some are just plain funny. Here are some examples:

Charles Darwin - *"If it could be demonstrated that any complex organ existed, which could not possibly have been formed by numerous, successive, slight modifications, my theory would absolutely break down. But I can find no such case."*

Richard Dawkins – *"The theory of evolution by cumulative natural selection is the only theory we know of that is in principle capable of explaining the existence of organized complexity."*

E.O. Wilson - *"Real biologists who actually do the research will tell you that they almost never find a phenomenon, no matter how odd or irrelevant it looks when they first see it, that doesn't prove to serve a function. The outcome itself may be due to small accidents of evolution."*

William Jennings Bryan - *"All the ills from which America suffers can be traced to the teaching of evolution."*

Tom DeLay - *"The causes of youth violence are working parents who put their kids into daycare, the teaching of evolution in the schools, and working mothers who take birth control pills."*

Mark Twain – *"What is Man? Man is a noisome bacillus whom Our Heavenly Father created because he was disappointed in the monkey."*

Milton Berle - *"If evolution really works, how come mothers only have two hands?"*

CHAPTER 5

THE SAGA OF MAN

"It was evident that such facts as these, as well as many others, could only be explained on the supposition that species gradually become modified; and the subject haunted me." - Charles Darwin (English naturalist, geologist, and biologist)

Let us begin by acknowledging that what is about to be described is widely accepted by the vast majority of the scientific community. Bear in mind that there can be a certain degree of interpretation and disagreement within the scientific community regarding any theory. As archeologists continue to unearth fossilized remains, and geologists and geneticists continue to do research, some details of the story of human evolution inevitably change. However, the principles that govern human evolution are no different than those that apply to every other species on the planet, and the main gist of human evolution remains broadly embraced. Also, bear in mind that *The Paleo Perspective* does not represent a detailed account of the fossilized evidence for the evolution of Homo sapiens. It merely highlights and summarizes some of the better-known facts. (The reader can easily Google anything that is being presented.)

It should also be noted that the few examples of our human ancestors presented here are not to be thought of as representing a linear model of human evolution. Indeed, human evolution is filled with dead ends (extinctions) and overlapping species—species that existed at the same time and place.

A case in point is a little village in Siberia called Soloneshnoye. Here the bones of three different types of early humans—Neanderthal, Denisovan, and Homo sapiens—have been discovered. The genome from the finger of a female child sequenced in 2010 revealed that Denisovans mated with modern humans, though not necessarily in this locale. Most modern humans also contain traces of Neanderthal DNA. Human evolution is complicated and certainly not linear.

Our Human Ancestors

Just who do scientists think some of our ancestors were? Among all living species, humans are most genetically similar to chimpanzees. On average, the DNA of humans and chimps is about 98.8% identical. This is no coincidence. Scientists believe that humans evolved from the same ancient ape-like creature as chimps did. In fact, biologists have been able to identify and count the number of changes, or mutations, that have occurred since the two species split apart. By analyzing these mutations, researchers have estimated that the last common ancestor of chimps and humans lived roughly seven million years ago. This is different than saying that humans evolved from chimpanzees, which is a common misunderstanding that many people share. What is being said is that humans and chimps both evolved from the same ape-like ancestor species, one branching off to eventually become modern-day chimpanzees, and the other branching off to eventually become modern-day humans. We did not evolve from today's modern-day chimpanzee.

Climate Change

As with many examples of evolutionary change, a change in the environment (a change in climate, for example) often precipitates the need for change within a species.

Around 10 million years ago, the climate in Africa began to change. Because of reduced rainfall, regions that had once been lush tropical forests became more arid, changing them from dense jungle to woodlands and savannahs. And so our ancestors had to become less dependent on trees for food and shelter, and more accustomed to moving about on the ground. Our human ancestors had to adjust—to evolve—to deal with this environmental change. Please remember that evolutionary changes happen very slowly. It took humans roughly 7 million years to evolve from our ape-like ancestor to modern man.

Here are a few examples of our ancient predecessors that are illustrative of our transformation from ape-like to fully human. Remember, this is by no means a comprehensive inventory; human development is quite complicated, with many prototypes along the way. Presented here are several of the better-known examples to give the reader a flavor of the evolution of our species. Scientists are making new discoveries all the time. Also, bear in mind that any pictures used here are just scientists' best guess as to what these hominins actually looked like. Nobody was around with a camera. (The term "hominin" usually refers to ancient relatives of humans that are closer genetically to humans than chimpanzees.)

Sahelanthropus

Scientists reason that the ability to walk upright would have been a great advantage in the aforementioned changing landscape. We now know that "hominids" (great apes and hominins) lived in Africa around six or seven million years ago. A skull found in north-central Africa represents a species named Sahelanthropus that probably existed at least 6 million years ago. Computer analysis of the fossilized Sahelanthropus skull suggests that the skull was balanced on top of an erect body, a trait that is unique to species that move upright. They probably walked upright when on the ground, but still spent much of their time in trees.

Ardipithecus

Ardipithecus lived about five million years ago and had a small brain measuring between 300 and 350 cubic centimeters. This is roughly the size of a chimpanzee brain. By comparison, human brains average around 1250 cubic centimeters. Ardipithecus was probably bipedal, as evidenced by its bowl shaped pelvis, the angle of its foramen magnum (where the spine enters the skull), and its thinner wrist bones. However, its feet were still adapted for grasping rather than walking for long distances. Its canine teeth were smaller than those of a modern chimp, but larger than those of modern humans. Scientists think that the reduced size of the upper canine teeth might be indicative of less aggressive behavior.

Australopithecus

Moving forward in time to somewhere between 4 million and 2 million years ago, we encounter several examples that can be grouped together under the heading/name "Australopithecus." There is evidence of at least 11 variations of this type of hominin whose existence overlapped during this time period. Because of the variety and complexity of the many fossilized specimens, and for the sake of simplicity, let us generalize and summarize what we know about this group of hominins.

Australopithecus had a rather prognathic face (protruding jaw line), with a flat nose, and a bony ridge over the eyes. Its large jaw contained canines and molars a bit more similar to modern man than to both modern and extinct apes. It had a rather low forehead, and had a brain size that ranged from about 380 to 430 cubic centimeters. Males averaged about 4 ½ ft tall, and were about 6 inches taller than females. The males weighed about 110 pounds compared to about 70 pounds for the females. Their strong upper bodies would have been useful for tree climbing. However, the femurs, pelvic bones, and the centralized position of the foramen magnum all indicate considerable upright, bipedal locomotion. Fossilized footprints look strikingly human and help confirm our belief that they did a lot of upright walking.

Additionally, evidence of stone tool use goes back somewhere to about 2.6 million years ago. Somewhere around this time, hominins were using flaked stone tools to both cut and scrape the meat from scavenged prey and to crack animal bones to get at the nutrient rich marrow contained within. Any hunting that might have been done would be made easier and more effective with a certain degree of cooperation and communication between the related individuals within a typical hominin troupe. Fossilized dental remains indicate that their diet also contained fruits, vegetables, and tubers.

Homo Naledi

Evidence of a new species of hominin was recently discovered (2013) in a cave named Rising Star, just outside of Johannesburg, South Africa. Named Homo Naledi, this species stood about 5 feet tall and weighed about 100 pounds. It had a brain size of about 500 cubic centimeters and teeth that appear to be more modern than that of Australopithecus. Homo Naledi had proportionately long legs and had feet that seemed fitted for walking great distances. However, its shoulders and hands suggested considerable climbing ability. As scientists continue to find more fossils in the area and in this cave, much more will be learned about this species. Preliminary estimates indicate that this species existed at least 2 million years ago.

Homo Erectus

As we take a look at another human ancestor, bear in mind that the "human tree" does not consist of a single branch, but instead, many branches. There are many examples of hominins that were around for long periods of time—and not just in Africa. Various hominins spread out of Africa and into Europe and Asia, with many different species existing in overlapping periods of time. In this sense the "human tree" is more like a bush. It seems that we have had lots of ancient "cousins," so permit me once again to generalize and summarize for the sake of simplicity and brevity.

"Homo erectus" refers to a type of hominin that existed as early as 1.8 million years ago, and as recently as perhaps 150 thousand years ago. However, most fossilized specimens are over a million years old. The designation implies a fully "upright" man. These hominins had long legs and arms. Females were about 5 feet tall, and males perhaps as tall as 6 feet, weighing 110 - 150 pounds. Early specimens had brain sizes as small as 600 cubic centimeters, and later varieties as large as 1100 cc. (Bear in mind that human brain size averages 1250cc.) Their faces were less prognathic, with smaller teeth than that of Australopithecus, and therefore more like modern humans. It should also be noted that many scientists figure that the loss of full body hair and the appearance of dark skin occurred about 1.2 million years ago. Chimpanzees, by contrast, have full body hair and light skin under their rather dark hair.

What was life like for Homo erectus? Evidence suggests that they were able to fabricate some fairly sophisticated stone tools, which they used for hunting and foraging. The commonly used term "hunter-gatherer" implies that their diet consisted of meat, nuts, berries, fruits, and tubers. Since other available evidence suggests that the habitual use of fire did not occur until about 300 – 400 thousands years ago, meat was probably eaten raw. (Raw meat is digestible.) Scientists figure that Homo erectus probably operated in small groups of perhaps 20 – 30 clan-related members, and there was most likely a high degree of social cooperation both at the base camps and while out hunting.

One thing that would have been very useful when a group (presumably mostly male) was out on the hunt would have been the ability to communicate and collaborate which each other. This would be especially true when hunting larger game or scaring away some rather large predatory cats. Additionally, the ability to trust each other to share the game, not only with co-hunters, but also with the women and children back at the camp, would be critical.

One should also picture the women sharing information about the location of various fruits, berries, nuts and tubers, and helping each other with childcare. Caring for each other's children would have allowed the women to be more productive and efficient with respect to the search for edible plant matter and with various other camp-related responsibilities. This pattern of trust and cooperation also requires that these hominins had the ability to delay gratification, understanding that cooperation in the present would lead to greater rewards in the future.

There is also some fossil evidence that suggests that this species cared for their sick or disabled cohorts. It is not difficult to conclude that these intelligent creatures understood that their individual survival depended on the survival of their group, and that surviving as an individual against the harsh unforgiving surroundings would be next to impossible. Man's ancestors were indeed highly social creatures. They had to be.

Homo Heidelbergensis

 Dates for Homo heidelbergensis range from 700,000 to 200,000 years ago. Evidence suggests that this early human species could be found not only in Africa, but in Europe and possibly Asia (China) as well. The term "heidelbergensis" was applied because initial discoveries were made in the vicinity of Heidelberg, Germany.

Heidelbergensis had a very large brow ridge, a larger braincase, and flatter face than earlier human species. It was probably the first human species to live in colder climates. Their short, wide bodies were likely an adaptation for conserving heat. Males averaged about 5 feet 9 inches tall and weighed approximately 136 pounds. Females averaged about 5 feet 2 inches tall and about 112 pounds.

Living in the colder climes of Europe and Asia, Homo heidelbergensis had the technical knowhow to control fire and fabricate wooden spears. He is thought to be the first human species to routinely hunt large animals. This early human is thought to be the first species to fabricate simple shelters using both wood and rock. The fact that he was building group shelters and participated in group hunting indicates a fairly high level of group cooperation and socialization. Most anthropologists are of the opinion that modern man (Homo sapiens) descended from Homo heidelbergensis.

Homo Neanderthalensis

Moving on to the more recent past, no description of human evolution would be complete without touching upon Neanderthals, or "Homo neanderthalensis." "Neanderthal" refers to the fact that the first fossil evidence of this ancient human was found in the Neander Valley in Germany. It is scientific consensus that Neanderthals also evolved from Homo Heidelbergensis, but is in fact not a direct linear ancestor to modern man. Apparently both Neanderthals and Homo sapiens have Heidelbergensis as a common ancestor.

Archeological evidence of Neanderthals suggests that they were around from about 300,000 years ago to as recently as 30,000 years ago. They ranged from Europe to parts of central Asia. There is some very recent evidence found in Israel (Misliya cave) that suggests that Homo sapiens migrated out of Africa at least 100,000 years ago (*Science Magazine 2018*). If that is accurate, there was a period of time in which Neanderthals and Homo sapiens mingled and could have interbred. To that end, researchers at the Plank Institute found that about 2.5% of the genome of the average *non-African* human is made up of Neanderthal DNA.

How would scientists describe our Neanderthal cousins? Neanderthals were a little shorter, but much stockier and sturdier than modern humans. Females were a little shy of 5 feet tall and weigh roughly 146 pounds. Males averaged about 5 ½ feet tall and weighed roughly 170 pounds. With strong arms, hands, and legs there is every reason to assume that they were athletic and as well coordinated as modern humans are today. These attributes would serve them well on the hunt. Their shorter, stockier bodies would be advantageous in the colder climes of northern Europe during the last glacial period. In spite of their physical prowess, life expectancy was only about 30 years (roughly the same as for Homo sapiens at the time).

What do their fossilized skulls tell us? Their average cranial capacity was 1600 cubic centimeters—larger than the 1250cc average for Homo sapiens. Neanderthals had a fairly large face, heavy brows, a rather large nose, and a receding chin. They had small molars but rather large incisors and canine teeth, hinting at the possibility that they used their teeth to do more work related tasks then just chewing. And although their cranium was larger, it had a slightly different shape than that of Homo sapiens, with more room in the back, and a slightly smaller frontal lobe region. The frontal lobe section is where complex critical thinking takes place, in addition to the capacity to engage in sophisticated social behaviors.

Scientists have theorized that Neanderthal behavior was less socially complex than that of Homo sapiens. For example, they traveled around in small, closely related groups (perhaps as small as ten), and in spite of the fact that these groups moved around a lot, covering large distances, there is a lack of evidence of any significant trading between the various clans.

Even though Neanderthals did not exhibit the capacity to socialize to the degree that modern humans do, they still led very complex, sophisticated lives. They controlled fire, cooked their food, constructed complex shelters, fabricated tools and weapons, and participated in organized hunts. Hunting was in fact one of their superior skills—routinely taking down horses, bison, deer, and rhino. There is no evidence so far that they used any sort of projectiles, such as a bow and arrow. Instead, they engaged big game with spear and axe in hand. You can imagine how dangerous this was, and their fossilized broken bones—similar to today's rodeo participants—prove it. Although meat was a large part of their diet, they also consumed some grains and legumes.

A significant amount of evidence suggests that Neanderthals buried their dead and cared for their sick and injured. There is a hint of evidence that there was some type of ritualistic behavior associated with these burials. A very tiny bit of evidence also suggests that this species fabricated some simple ornaments—such as a beaded necklace—and indulged in some body painting.

What about speech? Did Neanderthals have the capability to speak to one another? While there is no such thing as a fossilized "part of speech," there are several pieces of evidence leaving open the possibility of some form of vocalized communication. Researchers delving into the DNA of Neanderthal remains have found the gene crucial for the development of language—the FOXP2 gene. However, the presence of this gene does not alone guarantee that Neanderthals spoke to one other using anything that we would classify as language. Studies of their anatomy haven't answered this question definitively either. However, a bone in the Neanderthal throat, called the "hyoid," resembles that of modern humans.

We must consider Neanderthal to be a very successful species—after all, they were around for somewhere between 200,000 – 250,000 years. Why aren't they around today? Why did they go extinct? One theory is that there was some significant climate change in their habitat that greatly reduced the availability of big game, making it difficult for them to survive. Another idea is that competition with Homo sapiens proved too much, losing out to a species with superior skills—especially socialization skills. Still another idea is that Neanderthals were essentially overwhelmed and genetically absorbed by Homo sapiens. Even though this theory has its doubters, there is the previously mentioned evidence that all modern humans of *non-African* ancestry carry some Neanderthal genes. At any rate, we are here and Neanderthal is not. Let us finally take a good look at ourselves, Homo sapiens.

Homo Sapiens

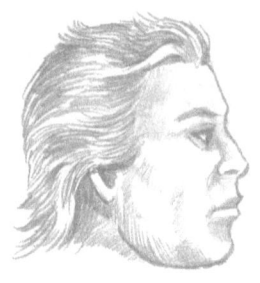

Let us take a good, scientifically objective look at Homo sapiens—after all, modern humans are the focus of this book. Admittedly, it is difficult for most of us to do so because of the proximity of the subject matter, and the alluring and egocentric idea that man is somehow "different." "Homo sapiens" literally means "wise man," although some might laugh at that label. Current data suggests that modern humans evolved from more archaic humans around 200, 000 years ago in Africa. There is some dispute as to whether our ancestors originated from East Africa or a little further south on the African continent. Some evidence seems to indicate that Homo sapiens began migrating out of Africa at least as early as 70,000 years ago—either heading north toward what is now Egypt, and then into the Middle East, or perhaps turning eastward and crossing onto the Arabian Peninsula. (There is some very recent evidence that this migration happened even earlier.) Either way, genetically modern humans spread out across the Middle East, into Europe, Asia, and eventually to the "New World" of North and South America. This represents tens of thousands of years of migration into different geographic and climatic environments. Most would agree that only a very clever, adaptable species could survive in all these different landscapes and climatic zones. Not only did we expand and survive, but we also multiplied to a present day population of over 7 billion.

Perhaps we should start with a physical description, just as we did with our ancient ancestors. Homo sapiens have a brain size that varies from about 1100cc to about 1450cc. The forehead rises sharply, eyebrow ridges are very small or non-existent, the chin is prominent, and the skeleton is somewhat gracile when compared to some recent ancestors.

Human body types vary substantially. Although genes play a big role in determining body size and shape, both are influenced by diet and exercise. The average height of a North American adult female is about 5 feet 4 inches and the average weight is about 137 pounds. The average height of a North American adult male is about 5 feet 9 inches and the average weight is about 172 pounds.

While female bodies are relatively hairless, some males, depending on their racial and ethnic background, have some thin hair on their chest and back. Current data suggests that humans lost most of their hair about one million years ago. There are at least two common theories why we lost our body hair. One suggestion is that a hairless body lends itself better to sweating and keeping cool across the hot African savannah. Another is that hairlessness is better for controlling lice, fleas, and ticks. Either way, we are for the most part a "naked ape."

Once our hair disappeared, dark skin became necessary as a protection from the intense African sun. Why aren't all human beings dark-skinned? One theory holds that once man migrated further north, protection from the sun became less important, but greater absorption of ultraviolet radiation by light skin helped the body synthesize more vitamin D. Consequently, the Homo sapiens that stayed in Africa remained dark skinned, while the Homo sapiens that migrated to the more northern latitudes evolved to have lighter skin color. However, we all started out as Africans, and we all started out with dark skin.

Humans have excellent color vision, although visual acuity in low light is limited. Our sense of smell, touch, and taste are moderately developed. Of all of the senses, the human ear is capable of responding to the widest range of stimuli. All five senses aid in man's outstanding ability to communicate with his fellow human beings. We are omnivorous animals that can consume both plant and animal products. However, it should be noted that Homo sapiens were originally hunter-gatherers, and probably consumed a decent amount of meat and fats whenever the opportunity presented itself. How many calories came from meat and how many came from vegetable matter is an unsettled question and probably depended on the availability of either.

Humans are capable of fully bipedal locomotion, thus leaving their arms and hands available for manipulating objects. The freeing up of the hands had important implications regarding tool fabrication and neurological development.

Sometimes early humans are referred to as Paleolithic man—in other words, "Stone Age" man. Early man fabricated many tools from stone. They included axes, choppers, scrapers, and knives. Sharply chipped stones were used as spear tips and arrow tips. Archaic tools also included bones and antlers, some of which were fashioned into needles for sewing animal skins. An example of a handy Paleolithic tool was the "atlatl." An atlatl is a simple but clever device that allows the individual to throw a spear faster and further than by hand. Tool fabrication is one of man's notable skills.

Paleolithic Times

What was life like 20,000, 50,000, 100,000, or 200,000 years ago? Archeological evidence and ethnographic parallels (studies of currently existing primitive tribes) suggest that ancient humans were nomadic or semi-nomadic hunter-gatherers (foragers). Man subsisted by hunting or trapping both big and small game, fishing, harvesting shellfish and insects, and foraging for fruits, vegetables, tubers, seeds, and nuts. Nature's abundance in any given area is finite, and so human groupings had to move whenever the local supply of food became exhausted. Possessions had to be limited to facilitate the necessary moving from one camp to the next.

Caves were often used as shelters, but some shelters were also fabricated from plant materials and animal skins. Some limited building materials may have been transported from camp to camp. There is archeological evidence that caves and man-made huts were communal in nature.

Human groupings had to be relatively small because only a limited number of people could congregate together without quickly exhausting the food resources of a particular locality. Such groups were typically comprised of either an extended family unit, or a number of related family units collected together in what we might call a "band." Group sizes ranged from as little as 20 to perhaps 50 or 60 members. It is reasonable to conclude that for these groups to operate effectively, a great deal of social cooperation and collaboration must have been present. Indeed, their survival depended on it.

Much, but not all, hunting was done as a group—probably mostly male. Hunting is much more effective when there is planning, collaboration, and coordination amongst the hunters, whether it's for bringing down large and dangerous prey or orchestrating drives that funnel small mammals or fish into nets or enclosures. Group hunting means that there must have been an expectation and a certain level of trust between the participants that the rewards would be shared amongst the hunters and brought back to camp for the women and children to share. Even when men hunted individually, there most likely was an expectation that the kill would be shared.

When women went out looking for vegetables, fruits, tubers, insects, or nuts sharing the bounty would also have been advantageous. Based on current ethnographic parallel studies, it is reasonable to assume that women would take turns watching each other's children so that more people could go out and scavenge.

It should also be noted that this day-to-day Paleolithic existence included man's use of language, and his penchant for art, music, dance, and religiosity—all human specialties. (A more detailed discussion on language, art, music, dance, and spirituality will be presented in a later chapter.)

This sharing, trusting, coordinating, and cooperating were all part of a kind of social contract—one that greatly enhanced the survival prospects of not only the group as a whole, but also the individuals within the group. Man is not the biggest or strongest animal on the planet, but his intelligence and level of socialization more than make up for his lack of physical prowess. As stated before, humans are the most socially complex animal species on the planet, and we possess some very special talents and abilities that help us maintain this very high level of social complexity. In the following chapter we will take a look at several examples of both non-human and human sociality.

Let us conclude this discussion by considering a few last points. First of all, we have to remember that organic life began on this planet over 3 billion years ago, and has been evolving ever since. Human evolution, per se, has been going on for roughly 6 or 7 million years, and little by little we became less "ape-like" and more human-like, both physically and mentally. Homo sapiens have been in the making for a very long time, and evolution has made us what we are today.

Secondly, the theory of evolution does not represent or express the notion that there is some sort of goal to evolution or a predestined end product. We should be careful about using the word "progress" to describe evolutionary changes. We mustn't think that evolution has led to perfection in any species—especially mankind.

For instance, humans have the capacity to get addicted to and crave drugs, alcohol, or sugar. We can also fall prey to destructive and compulsive behavior patterns like gambling or overeating. Any species' evolutionary design can possess some flaws. It just needs to be good enough to facilitate survival. All life forms, including us, for better or worse, are the result of evolutionary changes—adaptations to changing environmental conditions—nothing more, nothing less. Evolution facilitates survival, not perfection.

Different, but Not Separate

As pointed out earlier, it is very tempting to consider ourselves (Homo sapiens) to be different and separate from the rest of the animal kingdom. We often look for that one thing—that special thing—that makes us different. We are indeed different, but different in many ways, not just in a singular, "special" way.

Indeed, we are significantly different from our ancient Paleolithic ancestors like Australopithecus, Homo erectus, and Homo neanderthalensis. Our intellect, language, art, music, dance, culture, and spirituality help define what it means to be fully human. All of our special talents and traits have been essential for our success as a species. But it is a mistake to try to disassociate ourselves from our past, to ignore our genetic inheritance, and forget how we came to be. To understand modern man, we cannot ignore Paleolithic man. We are one and the same. This is the *paleo perspective*.

"Our own genomes carry the story of evolution, written in DNA, the language of molecular genetics, and the narrative is unmistakable."
- Kenneth R. Miller (American molecular biologist)

CHAPTER 6

THE CASE FOR HUMAN SOCIALITY

*"In the long history of humankind (and animal kind, too) those who
learned to collaborate and improvise most effectively have prevailed."*
— *Charles Darwin*

Non-Human Social Behavior

African Wild Dog

The Paleo Perspective advances the notion that man has evolved to be
a highly social and cooperative animal, and that this socially cooperative
group behavior is in large part responsible for his success. However, social
cooperation is not new or unique to Homo sapiens. Many present day
species of animals are also socially cooperative.

One such species is the African wild dog. These carnivores live in
closely-knit packs in which lone dogs are rare. Aggression amongst members
is muted, and cooperation is the key to survival and reproduction. They
are one of Africa's most successful hunters, a fact that has been attributed
largely to their high degree of cooperation.

Preceding the hunt, pack members often rally in a greeting ceremony.
During this ceremony social interactions include muzzle-to-muzzle
contact, lip licking, and lip biting. Different kinds of sounds can be
heard including whines, whimpers, squeals, and high-pitched bird-like

sounds called "twitters." Packs tend to split into groups during hunts, and eventually the separated pack members use a call to reunite.

Other examples of cooperation include the fact that they often leave kills behind for other members to eat, pups will be taken care of by dogs not on the hunt, and the adults allow the pups to eat first. All of this cooperative, social behavior has helped African wild dogs to be a highly successful species.

Honeybee

Let us look at another example of social cooperation—this one in the insect world. Honeybees live in complex societies that are referred to as "eusocial." Eusociality is a form of social behavior that includes the presence of several generations in a single nest at the same time, the caring for other member's offspring, and a division of labor among its members.

Any insect that lives in hives has a highly developed class system in which different members have distinct roles. Member roles for the honeybee include communication, food gathering, hive fabrication, and caring for the young. The queen bee directs the worker bees, produces offspring, and is in charge of food distribution. Worker bees, which are also female but are unable to mate, perform most of the work around the hive. They gather food, care for larvae, and guard the hive. The job of the male bees, called drones, is to produce offspring by mating with the queen. Each class of honeybee has a specific role to play in what is a highly organized, cooperative, and socially complex society. We should also note that many other successful species of insects (like the ant) are also highly social.

Chimpanzee

Humans come from a long line of socially adept animals. Ancestral primates have been around for approximately 80 million years, monkeys for about 30 million years, apes for about 14 million years, and chimpanzees for about 8 million years. Before we explore human sociality, let us look at some aspects of the social behavior of chimpanzees, with whom we share over 98% of our genetic makeup. (The following information can be found by accessing the Jane Goodall Institute of Canada and the National Primate Research Center, University of Wisconsin.)

Chimpanzees live in "fission-fusion" social groups consisting of a large community (up to several hundred) that includes all individuals that regularly associate with one another. There can be smaller, temporary subgroups, or parties. Membership in these subgroups—consisting of up to ten members—is unpredictable, and can be highly fluid. Subgroup membership can change very quickly, or it may last a few days. Eventually, the individual chimpanzee rejoins the community at large. Contact between members of these various scattered groups is accomplished by means of the distance call, the "pant hoot." At times the entire larger community will come together in large excited gatherings, usually when an abundant food source has been located. Primate behavioral ecologists have long debated the costs and benefits of group living. However, some of the possible benefits include decreased likelihood of predation, resource defense, feeding efficiency, and higher copulatory success.

Chimpanzee family bonds are very strong, especially mother-daughter bonds. Mothers and dependent young—up to age seven or so—are always together. Affectionate bonds also develop between family members and other individuals of the community at large. These can sometimes last a lifetime.

Chimp communities are male dominated, with age being an important factor. An alpha-male is usually between 20 and 26 years old. Some other factors that determine dominance and social status are physical fitness, aggressiveness, fighting skills, ability to form coalitions, and intelligence. Since there is a hierarchical system within chimpanzee societies, most disputes within a community can be solved by threats rather than actual attacks. Chimps use gestures and postures to demonstrate threat. Gestures include tipping the head, making hitting gestures, flapping hands in the air, swaying undergrowth, throwing objects, and charging towards one another. These gestures are often combined with vocalizations.

Studies suggest that male chimpanzee cooperation and altruism is not primarily and exclusively based on kinship. Examples include caring for unrelated orphans, mate guarding, group hunting, and meat sharing. Chimpanzee hunting probably evolved because of the direct benefits of a protein source in their largely frugivorous diets. Hunting is cooperative in the sense that multiple males are involved in cornering and capturing prey. Members of the hunting party spread out on the ground and in trees and vocalize throughout the pursuit.

The males of a community regularly patrol their boundaries, and if they encounter individuals of a neighboring community, they may attack with extreme brutality. Males sometimes form coalitions with each other to support each other during conflicts with other groups.

Visual and vocal communications are important in chimpanzee society. A host of facial expressions, postures, and sounds all function as signals during interactions between individuals and groups. Chimpanzees have particularly expressive, hairless faces. Facial expressions play an important role in close-up communication between individuals. For example, a "full closed grin" is in response to an unexpected and frightening stimulus, and evokes an instant fear response in other chimps. Other facial expressions include the "lip flip," "pout," "sneer," and "compressed-lips face." Vocal communications include "pant-grunts" (associated with dominance), the "bark" (associated with hunting), and the "tonal bark" (associated with the presence of large snakes).

Body position and stance also convey information to other individuals. Submissive positions include extending the hand, crouching, and bobbing. Aggressive positions usually involve an individual trying to appear larger than he is by swaggering bipedally, hunching his shoulders, and waving his arms. For a dramatic display, adult chimpanzees will also drum on the trunks of trees with their hands and feet. Drumming can be used in several other situations such as encountering other chimpanzee communities or when arriving at large food sources. Vocalizations, facial expressions, and body language are all essential for chimp communication, cooperation, and group functioning.

The African wild dog, the honeybee, and the chimpanzee are just a few examples of socialization in the animal kingdom. All animals, including humans, have an instinct to survive as individuals. Species that exhibit a good degree of sociality understand that their survival also depends on group cooperation.

Human Social Behavior

In an earlier chapter we discussed some of the better-known ancestors of Homo sapiens: Australopithecus, Homo erectus, and Homo neanderthalensis. Some time was spent on their physical characteristics as well as some on their inferred behaviors. Conclusions regarding their levels of socialization and cooperation were offered. Of course, knowledge of early man's socialization is difficult to ascertain, due to the fact that no one was around to record or film various early behaviors.

However, scientists do have two sources of evidence at their disposal to corroborate their findings and draw conclusions. Besides using whatever physical archeological evidence is available, many times scientists make use of what is called "ethnographic parallels." Ethnographic parallels are when we study relatively recent or even present day hunter-gatherer peoples to draw conclusions about the behaviors of hunter-gatherers of the past. Many studies have been done on Australian Aborigines, the San people (the Bushmen of Southern Africa), and Native Americans. And so physical archeological evidence is used in combination with ethnographic studies to draw reasonable conclusions regarding the behaviors of early man.

The chimpanzee is our closest living relative. The last common ancestor that we shared with the chimpanzee lived approximately six million years ago. There has been considerable physical and social evolution by our human ancestors over that period of time. We have learned to walk upright and lost most of our hair. We have multiplied and populated much of the Earth and invented many marvelous things. We have sent people into outer space and landed a few of them on the moon.

Considering these accomplishments, a strong case can be made that we should consider ourselves to be a successful species. But could we have come this far without a great deal of socialization—without a great deal of communication, cooperation, and collaboration? What is the evidence of early man's socialization?

A Successful Species

How do we assess the success of a particular species? One criterion to consider is the total worldwide population of said species. Generally speaking, the more the better. Another measure is whether the species can be found on most of the Earth's habitable continents—that is, whether it can thrive in varied environmental conditions. Still another consideration is over how long a period of time the species has managed to survive.

Evidence for the presence of ancient humans is found not only in Africa, but in Europe, the Levant, Central Asia, and Southeast Asia as well. And humans (genus homo) have been around for quite some time. Evidence suggests that anatomically modern humans—Homo sapiens—have been around for at least 200,000 years and have increased in number to over 7 billion.

Climatically, the Earth has gone through a series of warm periods and ice ages. Our ancestors lived and endured in all sorts of varied geographic and climatic conditions, and did so over vast periods of time. This is also a measure of our success.

Previously mentioned is the fact that many species rely on sociality to enhance success. Indeed, the effectiveness of sociality is argued quite convincingly in Edward Wilson's book *The Social Conquest of Earth* (2012). And humankind is a case in point. So let us look at examples and evidence that attests to human socialization—man's proclivity for effective group behavior.

Dissemination of Lithic Technology

The ability to fashion stone tools goes back over 2 million years. The ubiquity of these various stone tools would not have been possible without the successful transfer and dissemination of this lithic technology from individual to individual, from group to group, from location to location, and from generation to generation. The transfer of this kind of knowledge is not possible without effective communication and collaboration amongst group members. It goes without saying how important tool fabrication was for human groupings, and one cannot imagine our ancestors not wanting to share all the various stone tool fabrication techniques with each other.

The Controlled Use of Fire

There is considerable evidence that early hominins knew how to control and make use of fire several hundred thousand years ago. There is even some speculation that the use of fire may go back to over 1 million years ago—but the evidence for this is somewhat scant. What scientists do find are the remains of fireplaces and hearth like depressions containing burnt bones and stone tools—most likely used for the butchering of prey.

For example, there is evidence (burned seeds and wood) of fire use at The Gesher Benot Ya'agov in Israel that dates back almost 800,000 years. There is also evidence of early man heating up silcrete stones (silica cement) to make them more workable before knapping. (Knapping is the striking of a core stone to shape it into a stone tool.)

The communication and transfer of the knowledge necessary to start a fire, maintain a fire, use it to cook food, or use it to help make better stone tools are also examples of effective group behavior.

Hunting Together

Another example of hominin sociality and group functionality can be found in their hunting habits. There is much evidence of large kills. Remains of horses, bison, rhino, and elephant have been discovered. It would be highly unlikely that a single hunter could manage to bring down such large and formidable prey. It is more likely that ancient man hunted in organized groups, thereby increasing the likelihood of success. Using various ethnographic studies, we can conclude that the kill would have been shared amongst the hunters, and that it would have been brought back to the camp to be shared with the rest of the troupe. Individuals who did not directly participate in the hunt were most likely engaged in searching for edible plant matter such as berries, fruits, and tubers. The sharing of all of these foods was essential for group survival.

There is also the fact that group hunting and foraging requires individuals to be away from the base camp for extended periods of time. This means that at times children would have to be looked after by individuals other than their parents. And so childcare would also have had to be a shared, group experience. Furthermore, all this cooperation and collaboration requires a healthy degree of trust, and trust is an essential ingredient for positive group dynamics. This cooperation, collaboration, and sharing amongst the members of a species is sometimes referred to as "reciprocal altruism."

Living Together

There is lots of evidence of early man using caves for shelter. These caves were large enough to house multiple family units, contained evidence of multiple campfires/hearths, and included the remains of a significant number of cooked animals—too much for one family.

One example of a rather large cave is Border Cave in the Lebombo Mountains in KwaZulu-Natal (South Africa). This cave was over 100 feet across and appears to have been continually used by many generations of humans for over 100,000 years. Another such example is Spain's Abric Romani cave (approximately 50,000 years old) which had multiple levels, and in which there is evidence of 16 hearths on just one of the several levels.

There are also several examples of multiple-family *manmade* structures. One such structure is an "Acheulean" hut found in the Grotte du Lazaret, near Nice, France. This tent-like structure was built inside a cave and was made of animal hides draped over a wooden framework. It dates to somewhere between 400,000 and 500,000 years ago. The interior measures about 33 feet by 11 feet and was subdivided into several rooms.

Another archeological site, at Kostienki, near Alexandrova in the Ukraine, has evidence of a manmade hut measuring approximately 100 feet by 17 feet. This site dates back to about 40,000 years ago. Most of these ancient huts were constructed from wood and animal bones.

Still another example of group living can be found at the archeological site at Dolni Vestonice, Morovia (Czech Republic). This site dates back about 25,000 years and has a hut containing five hearths.

A more recent example of group habitation can be found at CatalHoyuk in southern Turkey. Here the houses were made of mud brick and were in an apartment-like configuration. The units adjoined each other, similar to cells in a honeycomb. The site covers about 32 acres and probably housed thousands of people. The older, bottom layer of buildings dates back to approximately 7,500 B.C.E., and the newer, top layer of buildings dates back to about 5,600 B.C.E.

There are many more examples of large, ancient, multifamily habitats. Whether it was caves, wood and bone huts, or mud brick domiciles, ancient man was clearly a social dweller. Living together in groups ranging from several dozen to over a thousand requires an enormous degree of social cooperation.

Caring for the Sick

Another indicator of man's social proclivities is evident by his willingness to care for sick or disabled group members. Evidence for this kind of altruistic behavior goes back over one million years. KNM-ER 1808 is the remains of an adult female from the Koobi Fora site in Kenya, and is about 1.5 million years old. Indications are that she suffered from hypervitaminosis–A. This disease causes abdominal pain, nausea, headaches, dizziness, blurred vision, a reduction in bone density, and a lack of muscular coordination. For this individual to make it to adulthood, a great deal of assistance and care from the other members of her group would definitely have been required.

Another example of human kindness is illustrated by the remains of a 500,000-year-old specimen from the Sima de los Huesos site in Atapuerca, Spain. This child had craniosynostosis, which caused the subject to have a deformed face and head, brain damage, and a high probability of mental retardation. Scientists estimate the age of the individual at death to be about ten years old. Clearly this child would not have survived that long without care from the surrounding adult members of his or her (undetermined sex) group.

A third example of human compassion comes from Shanidar cave in Iraq. This site dates to approximately 70,000 years ago, and provides evidence of an adult that sustained a fracture in the upper right arm. This injury most likely caused the arm to atrophy, and thereby render the arm to be virtually useless. This individual also had a crippled right leg and a fractured foot. It would be hard to imagine that this person could have made it to adulthood without considerable care and help from other members of his group.

One last example comes again from the Dolni Vestonice site in the Czech Republic. This site is about 25,000 years old and contains the remains of a cared-for female with chondrodysplasia calcificans punctata. This disease causes defective bone growth, coronal clefting, and joint contractures. It is sometimes accompanied by seizures and mental retardation. The subject also had additional fractures of her upper limbs.

All of these are examples of altruism—caring that goes back over one million years. They are indications of the high degree of group cooperation and socialization that had been developing for quite some time in our species.

Bigger Brains

There is one last thing that needs to be mentioned as evidence of man's ability to socialize. It is the one thing that makes all this complex socialization possible. Tool fabrication, group hunting, group living, and the use of language—none of these would be possible without a rather large and sophisticated brain.

However, when we talk about brain size we must consider several critical factors. The first is the relative size of the brain compared to body size. The second is relative size of the "neocortex" compared to the other parts of the brain.

The human brain has a neocortex that is relatively large compared to other parts of the human brain. For example, the human neocortex is over one hundred times larger than the "medulla." The medulla is responsible for some basic bodily functions such as breathing, heart rate, and blood pressure. The neocortex is believed to be the seat of some higher cognitive functions such as reasoning, mathematic ability, language, and something called "social intelligence."

While humans do not have as big a brain as some other animals—elephants, for example—we do have a higher brain to body weight ratio, and have a higher degree of social intelligence. This social intelligence allows humans to lead a more socially complex life than do elephants.

Social Intelligence

Human social intelligence refers to our ability to handle socially complex scenarios such as friendships, romance, deception, reciprocity, and altruism. It is believed that the evolutionary need for complex human socialization was a major driving force in developing the size and complexity of the human brain. It was the demands of living together, hunting together, cooperating, and collaborating with each other that drove our need for social intelligence.

Archeologist Steve Mithen believes that there are two key periods of human brain growth that are relevant to the development of human social intelligence. The first was around 2 million years ago, when the brain more than doubled in size, from about 450cc to 1,000cc. Mithen believes that this growth was necessary because people were living in larger, more complex groups. They had to keep track of more people and more relationships, which required a larger brain.

The second growth in human brain size occurred between 600,000 and 200,000 years ago, when the brain reached its modern size. Mithen believes that this second growth spurt was related to the evolution of language. Language is probably the most complex cognitive task we undertake. It is directly related to social intelligence because we use language to navigate through and mediate our social relationships. Some specific examples would include listening, negotiation, conflict avoidance, and conflict resolution. It is perhaps an understatement to say that humans rely heavily on their social skills to facilitate group survival. (There will be a more detailed discussion of language in a subsequent chapter.)

"Individual commitment to a group effort—that is what makes a team work, a company work, a society work, a civilization work."
- Vince Lombardi (American football player and coach)

CHAPTER 7

GENETICS AND HUMAN NATURE

"DNA is like a computer program but far, far more complicated than any software ever created."
– Bill Gates (American business magnate and co-founder of Microsoft)

As mentioned before, a *paleo perspective* is a scientific perspective, and one that is centered on our genetic makeup. This perspective helps create a mindset that not only affects how we look at the world around us, but how we analyze human nature itself.

In subsequent chapters, we will be discussing various aspects of human nature. That being the case, it would be helpful to have a rudimentary grasp of genetics and how our genetic makeup influences our behavior. You may already know the basics if you have taken biology in high school or college. If you have not, I hope this brief description is helpful and at least gives you a feel for the role that genes play in our lives. Bear in mind that *The Paleo Perspective* is not a biology textbook, and so the following description is both introductory and elementary. All of this information is readily available in any basic biology textbook and easily accessed on the Internet.

Genetics 101 - "Abbreviated"

Most people have heard of "DNA," which stands for "deoxyribonucleic acid." The DNA molecule is an extremely long, narrow, and complicated molecule. It contains over 200 billion atoms. If you were to stretch it out it would be about six feet long, but this string-like molecule can be twisted and coiled up and wind up measuring only about .005 millimeters. DNA is found in a special area of the cell called the "nucleus." Because the cell is very small, and because organisms have many DNA molecules in each cell, each DNA molecule must be tightly coiled and wound to fit in the tiny nucleus. This tightly packaged form of the DNA molecule is called a "chromosome." Human beings have 23 different chromosomes.

Chromosomes differ from one another in size and function. A particular section of DNA located at some particular spot on the chromosome is called a "gene." There are many such genes on each chromosome, and each gene contains a different piece of DNA with a different set of instructions. One gene may influence hair color. Another may affect a person's height. Still another may contribute to a person's level of musical talent.

Scientists use the term "double helix" to describe DNA's winding two-stranded chemical structure. This shape, which looks like a twisted ladder with rungs, gives DNA the power to pass along biological instructions with great precision. The rungs of this ladder contain over 20,000 genes, each with a different set of instructions. The structure of the DNA molecule lends itself easily to replication because the sides of the double helix run in the opposite (anti-parallel) directions, and can be unzipped down the middle. Each side can then serve as a replication template for the other side. Please observe in the following picture a cell with its nucleus, a chromosome from the nucleus, the DNA helix, and a section (gene) on the DNA strand. Note that the chromosome is the coiled configuration of the DNA.

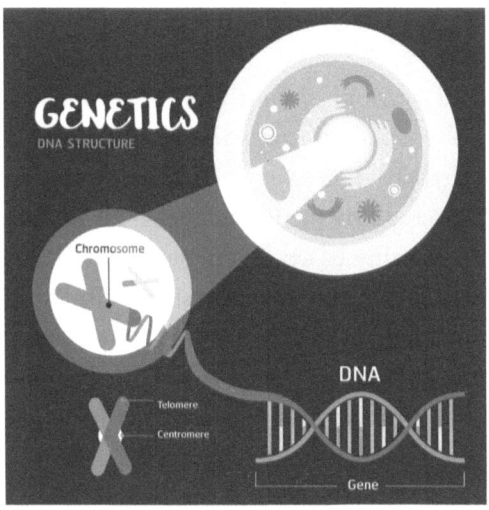

Suffice it to say that in all living things it is the job of the DNA to direct the replication of its various cells, and to facilitate the replication of the organism as a whole. In asexual reproduction there are no male and female versions of the species, and so the asexual organism reproduces itself. In sexual reproduction—as is the case with Homo sapiens—there is a male and a female version of the organism, and reproduction requires that genetic material from the male combine with genetic material from the female to produce offspring.

During sexual reproduction there are several stages where genetic material from the male and female are combined, swapped, and rearranged. This produces an offspring that is not only genetically different from both parents, but also contains genetic splices from both the maternal and paternal grandparents. When egg and sperm meet, the baby inherits a combination of genes that is totally unique. It carries versions of genes from all 4 grandparents, plus any mutations (mistakes) that occurred when the egg and sperm were made by the mother and father.

The intricacy and complexity of this process, which involves the mixing and reordering of thousands of genetic bits, ensures genetic variety in the offspring. As mentioned in a previous chapter, this genetic variation in the offspring is essential for the process of natural selection and the process of evolution in general.

Genes and Your Behavior

As we have said, sections of your DNA, called genes, are responsible for your various physical characteristics. But your genes have influence over many of your behaviors as well. Some of these behaviors can be categorized as instinctual. By instinctual behavior we mean behavior that is performed without being based on any prior experience (in the absence of any learning), and is therefore the expression of innate biological factors. Examples would include a baby's instinct to cry, our instinctual avoidance of bitter foods, and a mother's instinct to nurture her young. Some might be surprised that even our selection of a mate is influenced by genetic urges. Consider the following experiment performed in 1995.

> *In the first "sweaty T-shirt" experiment (1995), a Swiss zoologist, Claus Wedekind, set out to test a woman's sensitivity to male odors. He assembled 49 women and 44 men. He gave the men clean T-shirts to wear for two nights. After the two nights the shirts were then returned to the scientists.*
>
> *In the laboratory, the researchers put each T-shirt in a box equipped with a smelling hole and invited the women volunteers to come in and sniff the boxes. Their task was to describe each odor as to intensity, pleasantness, and sexiness.*
>
> *The results were striking. Overall, the women preferred the scents worn by men whose immune system was genetically different from their own. By selecting a mate that had an immune system that was genetically different from their own, they were enhancing their potential offspring's chances of having a more diverse immune system than either parent. This of course would increase the survival chances of their offspring.*

There are, of course, many other factors besides odor that affect the selection of a mate. There are both genetic and social factors at work. But your genes play a big role in many of your behaviors—more than many people may think.

In addition, genes can get "turned on" or "turned off" by both internal and external environmental factors. For example, gender-dependent hormones can affect gene expression. A case in point involves maternal milk production. Although the genes for producing milk are carried by both males and females, it is only after a woman gives birth that her pituitary gland produces the hormone "prolactin," which stimulates the production of milk.

Similarly, drugs, chemicals, temperature, and light are among the external environmental factors that can determine which genes are turned on and off, thereby influencing the way an organism develops and functions. An example of the way in which the ingestion of certain chemicals can alter gene expression involves thalidomide. Thalidomide, manufactured and marketed during the 1950's, was originally prescribed as a sedative to help alleviate anxiety and tension. While thalidomide has no discernable effect on gene expression and development in healthy adults, it has a profoundly detrimental effect on developing fetuses. Sadly, thalidomide caused severe developmental deformities in over 8,000 infants.

The Nature-Nurture Debate

Human behavior is quite complicated. Your genes do play an important role in your behavior, but so does the environment. Your behavior is influenced not only by what you have been taught at home, school, or a religious institution but also by what you may have learned on your own through personal experience.

How much of our behavior is innate (comes from our genes) and how much comes from what we have been taught? This age-old question is sometimes referred to as the "nature-nurture" dilemma. "Nature" refers to instinctual or innate behavior, and "nurture" refers to what you have learned from your immediate surroundings or have absorbed from society in general.

"Genes are deterministic but they are not destiny."
- Jennifer Ouellette (American author- mainstream science)

Many books have been written dealing with this question, both fiction and nonfiction. For instance, *Lord of the Flies* (William Golding, 1954) is a novel that deals with a group of youngsters who get stranded on a remote island. In the absence of the usual norms and constraints of civilized society, the boys resort to some rather primitive, instinctual behaviors. The book describes the boys resorting to ritualistic killing, chanting, dancing, and participating in a sacrificial offering to a fictitious beast that they imagined haunted the island.

Many times scientists look to the very young—infants and toddlers—to give us clues as to what is instinctual and what is the result of learning. This is based on the idea that for the very young there has been less time for their behavior to be modified by their parents' guidance or the surrounding society. Their behaviors would therefore be more *nature* and less *nurture* than corresponding adults.

Instead of trying to answer the nature-nurture question with some sort of mathematical formula (like 65% nature and 35% nurture), let us consider several analogies that will hopefully give us some insight into this complicated question. These analogies are meant to give us several easy-to-imagine examples—ones that most of us have had some experience with—to help us picture how our genes and the environment interact and combine to fashion and affect our behavior.

Alphabet Soup

Consider if you will a bowl of Campbell's brand alphabet soup. The soup contains a good supply of all the letters of the English alphabet. The letters can combine to form words, and the words can combine to form meaningful sentences. Our genes are kind of like those letters and words in the soup in that they can spell out instructions regarding both our physical characteristics and our behaviors.

For example, the letters in your "genetic soup" might spell out "tall," "thin," "blue eyed," "musically talented," or "shy." In our model this would be analogous to your genes possessing the chemical instructions for you to be tall, thin, blue eyed, or shy.

External events, such as stirring or heating the soup, might cause some letters and words in the soup to sink to the bottom and out of sight, while others might come to the surface, and change what the soup "has to say."

Similarly, environmental factors can reduce or diminish some genetic tendencies while enhancing or emphasizing others. Outside influences can bring some genetic propensities to the foreground and push others further down in that genetic soup. One particular genetic proclivity might overpower another. A particular genetic trait might never get to "come to the surface" and express itself in your genetic soup.

For instance, you might have the genetic potential to be a great musician, but your economic status or family situation might make it very unlikely that you would have the opportunity to study music and make a career out of it. Or, looking at this situation from the other side, you might be part of a well-to-do family, with lots of resources and opportunities to become a musician, but the genetic material necessary to become a really great musician is simply not in your genetic soup.

Consider another illustration of how nature and nurture interact. Suppose you have the genetic propensity for risk-taking behavior. Using our soup analogy, we might imagine that somewhere in your genetic soup your genes are spelling out "risk taker." Will you take a risk and experiment with cigarettes or maybe even some type of illegal drug? After all, the genetic potential is there.

Let us continue with our illustration by supposing that you live in a crime-ridden neighborhood, with easy access to illegal drugs, and many of your friends are users. Additionally, you are part of a poverty stricken, dysfunctional family, and attending a dangerous and underperforming public school. All of these environmental and social factors would add to that inborn "risk-taking" type message in your genetic soup and enhance the likelihood that you would engage in illegal and dangerous behavior.

However, if you came from a stable family, went to a really good school, lived in a safe neighborhood, and had drug-free friends then the outcome might be different. Your risk-taking gene might not be expressed in terms of some sort of illegal behavior, but instead might be expressed by you opening your own business.

Perhaps we can think that having positive environmental factors would be like adding sugar or tasty herbs to your genetic soup. This would have the effect of mitigating some of those bitter or spicy risk-taking genetic ingredients. A person can have the potential for risky or bad behavior, but positive environmental "additives" can mitigate some of those negative instincts. (For the reader with a more extensive science background, we would say that the environment affects "gene regulators," which in turn affects genes, which in turn affects behavior.)

We can use the soup analogy to think of a human being as having inherited a unique (everybody is different) "genetic soup," but as time goes on, external factors can add additional ingredients (messages) to that soup, changing the overall character of that soup. And so, a person's physical and behavioral characteristics will be a combination of both genetic and environmental factors. Perhaps it is best to think of it as "nature *and* nurture" as opposed to "nature *or* nurture."

Baking a Cake

Let us consider another analogy that might give us some additional insight into the nature-nurture quandary. Think about the baking of a cake. The proper ingredients need to be mixed in addition to stirring and heating. In this analogy the original, basic ingredients correspond to your genetic makeup. But we can alter the resulting cake by adding or subtracting ingredients, or by preparing it differently. These actions would be analogous to outside environmental and social factors. And so the original recipe, with the original ingredients, is not the final word on how that cake comes out—just like your original innate genetic profile is not the only factor in determining how you turn out. The fact that you may not be a naturally gifted math student doesn't mean you can't study hard, get some extra help from the teacher, and pass that math course. You might even go on to become a top-notch accountant.

There is a parable in the Christian New Testament where Mark (4:16 – 20) likens the word of God to seed that must fall on good soil to bear fruit. Perhaps we can liken the seed in this parable to the fact that sometimes good genes need a nurturing environment to see their potential expressed.

Conflicting Genetic Messages

Our genetic makeup is quite complex, and there are times when one genetic urge can be in conflict with another. As a case in point consider "maternal instinct." Most people agree that the vast majority of women have a biological urge to protect their offspring. There are times when a mother will even sacrifice her own life for her children. But this is not always the case. Occasionally, a woman's urge to survive, to save her own life, proves stronger than her maternal instincts. In these life and death situations your genetic urge to survive can be at odds with your parental instincts.

As we read the daily newspapers or watch the news on TV, we see instances of people committing all sorts of crimes. But we also read stories of individuals doing heroic and altruistic deeds. We see both extremes displayed routinely as life plays out around us. Sometimes we even see both good and evil within the same individual. When thinking about hardened criminals, many people say that even the worst of them have a little good down deep inside. It is also said that everybody has a breaking point, and that each of us is capable of evil given the most dire of circumstances. Human behavior is indeed quite complicated.

Nature has given each of us the genetic material to be both bad and good. And in most situations there is the opportunity for us to consciously decide to either summon up the good, or acquiesce to our baser instincts. This leads us to the concept of "free will."

Free Will

There is one more factor that we need to be mindful of when thinking about what things affect human behavior, and that is the concept of free will. The principle of free will has religious, legal, and scientific implications. Putting aside the question of exactly how our brains operate on an atomic, quantum mechanical level, most people agree that for the vast majority of situations, human beings exercise their will as they make decisions—minute by minute, day by day.

This is not to say that we always have a say in what our choices are in any given situation. Many times we have limited control over the many environmental and social circumstances in which we find ourselves. There are times when there are no good choices, or perhaps no choices at all. And none of us had any choice in what genetic qualities we inherited at birth. But given our genetic makeup, and given whatever the social and circumstantial conditions that we find ourselves in, we are still the final decision maker. We have the option to exercise our free will.

> *"Life is like a game of cards. The hand you are dealt is*
> *determinism; the way you play it is free will."*
> *- Jawaharlal Nehru (First prime minister of India.)*

The question of free will came to the fore as a reaction to, and criticism of, B.F. Skinner's theory of "operant conditioning." (Operant conditioning refers to the changing of behavior as a result of some type of reward or punishment.) During the early twentieth century, when the principles of behaviorism and operant conditioning were being developed, there was a negative reaction to the theory within some religious circles regarding the question of free will. Much of the work done by B.F. Skinner and other behaviorists was done in a laboratory setting using mice, rats and other types of lab animals. People wondered if man could be "conditioned" like a mouse, rat, or dog. Where is man's free will?

Skinner wrote several books, one of which was entitled *Beyond Freedom and Dignity* (1971). One of the key concepts developed in this book was the idea that once we discover, understand, and accept what makes us do some of the things that we do (as in operant conditioning), we are then free to exercise some control over various negative and unwanted behaviors. B.F. Skinner reasoned that ultimately man could exercise some control over his destiny. Once man understands and accepts the role that his genetic makeup plays in his life, and once he understands and accepts the environmental influences that affect him, not only is he then in a position to exercise his free will, but it is in fact man's responsibility to exercise that will.

For example, if you know that you have inherited a genetic propensity to be overweight, then it is up to you to pursue a lifestyle that will mitigate that innate predisposition. That might mean buying the right foods, being selective when dining out, joining Weight Watchers, and exercising on a regular basis. Furthermore, if you know that you tend to buy too much food—especially junk food—when you go to the supermarket hungry, then you should think about this proclivity before planning your next trip.

Similarly, if you like to drive fast, and already have several speeding tickets, why buy a high performance sports car? Or, if you like to gamble just a little too much you shouldn't pick Atlantic City as your next vacation destination.

Consider one more case in point. This involves man's propensity toward racial and ethnic prejudice. Once we understand and accept the fact that human beings have an innate tendency to be more suspicious of people that do not look or sound like they do, or have different customs than they do, then it is up to us to facilitate social situations that would mitigate those instinctively negative propensities. For instance, you might purposely expose yourself and your children to people of different cultures, races, religions, and ethnic backgrounds. Likewise, making an effort to ensure some diversity in neighborhood schools or in your community would be a step in the right direction.

These situations involve acknowledging and accepting some inherently problematic aspects of human nature and deliberately choosing to exercise your free will to produce a more desirable outcome. Many times these decisions are very difficult to make, and even more difficult to implement.

Genes, Environment, and Free Will

We can think of human behavior as the convergence of three factors: our genetic makeup, environmental circumstances, and free will. Just how much does each influence our behavior? Life is too complicated for us to assign specific mathematical percentages to these three variables in any given situation.

We can however get a feel for how these three factors interact by imagining the following analogy. This analogy involves a car with your "free will" sitting in the driver's seat. You might have been born an expensive, high-performance type of sports car. You've got all-wheel drive, lots of horsepower, rack and pinion steering, and a sophisticated suspension system. Or, you could have been born an "economy" type vehicle, with none of the above features—just the basics. In this example, these features are analogous to your inherited genetic attributes.

Now the road in front of you may be flat or hilly, straight or dangerously curved. It may be dry and sunny out, or it might be wet and slippery. Maybe there is a lot of traffic on the road, with lots of crazy, dangerous drivers cutting you off and driving you insane. These represent environmental factors.

Knowing which kind of vehicle you are is akin to self-awareness. How you get from point A to point B in those various road conditions is where your free will comes in. Do you drive safely, or do you take chances? The bottom line is that no matter what kind of car you are (your inherited genes), and no matter what kinds of social and environmental conditions you experience on the road of life, your free will is still in the driver's seat.

We all have inherited, innate attributes—some good and some not so good. *The Paleo Perspective* reminds us that some of these inherited attributes are thousands of years old. And we all are subject to various social and environmental forces—some wonderful and some horrific. But it is still up to us to do the best with what life has dealt us and with what life throws at us. It is incumbent upon us to exercise our free will. Indeed, it is our moral responsibility to do so.

"Virtue lies in our power, and similarly so does vice; because where it is in our power to act, it is also in our power not to act ..."
- Aristotle (Greek philosopher and scientist – 350 B.C.E.)

CHAPTER 8

OUR PALEOLITHIC BRAIN

"The x-ray of your skull shows a large, floppy mass floating
inside. I have to consult my colleagues to be certain, but
it looks like a long sausage snarled into a lump."
- Benson Bruno (American author and satirist)

Once again, let us proceed with a word of caution before we begin our discussion on the human brain. *The Paleo Perspective* is not meant to be a textbook-level description of the human brain. This chapter represents a somewhat superficial and easily verifiable description of the subject matter, but one that is necessary to employ a *paleo perspective*.

The Paleo Perspective considers a discussion on the human brain to be relevant for several reasons. The first is the fact that the evolution of the brain is a case in point regarding evolution in general. Changes in the brain—just like changes in physique—are the result of evolutionary pressures and natural selection.

Secondly, our brain directs much of our behavior. Understanding just how the brain operates helps us gain insight into human nature in general, and such particular issues as fantasy and myth genesis, worldview formulation, and something called "theory of mind." These are all key issues that this chapter will be viewing through a Paleolithic lens. Let us recall that the primary thesis of this book is that humans are operating with a 200,000 year-old design—our brains included.

The Ultimate Adaptation

The primary function of your brain is to enable you to survive and reproduce. In a way, the brain is carrying out the directives of our DNA. As stated earlier, our DNA is coded for survival. This genetic coding has the instructions to develop a brain—a brain whose main task is to maximize our survival prospects. In order to do that the brain has to take care of some basic bodily functions, like breathing and moving about, but it also must take care of higher order tasks, like problem solving.

One might consider man's highly evolved brain to be the ultimate adaptation. While other animals have developed things like a long neck (giraffe), big claws (tiger), or echolocation (bat) to help them survive, man—in addition to various physical modifications—has developed an amazing brain. We might even argue that the evolution of the human brain circumvents the necessity for various other physical evolutionary modifications. For example, by inventing the ladder, man has less of a need to evolve into a taller and taller species. By inventing the knife, man need not have such large and pointed teeth (fangs). And by inventing clothing, man did not need to develop a thick coat of hair to keep warm in colder climes.

This larger and more complex brain has enabled man to not only deal with all sorts of ecological and environmental changes, but has also enabled him to multiply, spread out, and occupy virtually every part of the planet. It is man's cognitive abilities that have propelled him to such a dominant position here on Earth. Homo sapiens find themselves at the very top of the food chain, and to quote Genesis (1:26): "And let them have *dominion over* the fish of the sea and over the birds of the air, the domestic *animals* all over the earth, and all the *animals* that crawl on the earth."

Brains 101 – "Abbreviated"

"The human brain has 100 billion neurons, each neuron connected to 10 thousand other neurons. Sitting on your shoulders is the most complicated object in the known universe." - Michio Kaku (theoretical physicist)

Our brain weighs about three pounds and contains about one hundred billion neurons (nerve cells). Each neuron links with about 10,000 other neurons to form an incredible three-dimensional grid. This means that the human brain has approximately a thousand trillion (1,000,000,000,000,000) neurological connections. That is an enormous amount of connectivity.

The brain can be thought of as having three layers—the oldest layer being the innermost layer, and the outer layer being the newest. The centermost part (the oldest part) is called the "brain stem" and is dedicated to survival functions. This part of the brain is sometimes referred to as the "reptilian" part. The middle layer is the "limbic" system, whose domain is our emotions. This part of the brain is sometimes referred to as the "mammalian" part. And the outermost layer (most recent) is the "neocortex," which is involved with higher order functions such as sensory perception, motor commands, spatial reasoning, conscious thought, and language.

Humans would like to think that their neocortex dominates their brain's functioning, but much of the brain is dedicated to instinctual behaviors. Sometimes people say that they are going with a "gut feeling," or they are following the wishes of their "heart instead of their head." These are attempts to explain instinctual behaviors that do not seem to originate from within the neocortex. Let us keep in mind that the human brain has some very ancient parts.

In addition to thinking of the brain as being a three-layered organ, we also know that specific areas of the neocortex have specific functions. We can observe from the following picture that there are areas dedicated to such functions as speech, vision, hearing, motor control, smell, and facial recognition. It is therefore helpful to think of our brains as operating in a "modular" fashion. This is in contrast to likening our brain to a computer with a central processing unit (CPU) through which all operations get handled. Our brains do not possess a CPU.

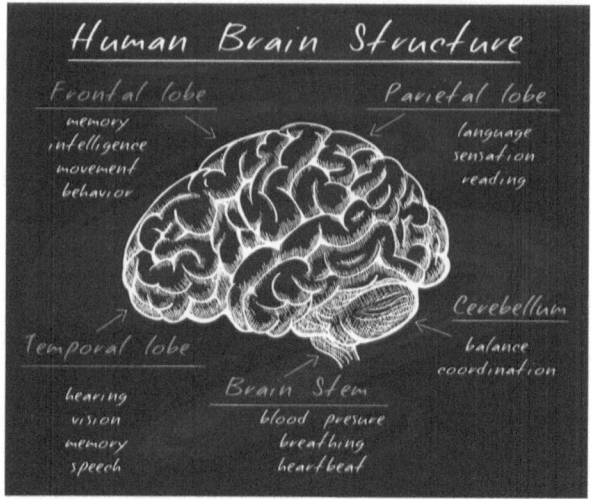

The Modular Theory of the Brain

The human brain differs significantly from a modern computer. A computer with a large, fast CPU would do all tasks faster and better than one with an inferior CPU. This is not the case with the human brain. It was not designed as a whole system "from the ground up," and is instead a conglomeration of "modules." Our brain evolved over a very long period of time, adding layer upon layer, section-by-section, module-by-module.

The concept of brain modularity is the idea that for the most part, our brain is composed of various clusters of neurons (modules) that have distinctly established functions that are task-specific and have evolved in response to the forces of natural selection. (Many of these ideas have their basis in the relatively new field of evolutionary psychology, pioneered by the work of anthropologist John Tooby and psychologist Leda Cosmides.)

The "modular theory" of our brain is critically important to help explain the evolutionary purpose of many specific human aptitudes such as facial and voice recognition, eye-hand coordination, art or music talent, and a sense of direction. Several of these aptitudes, such as voice and face recognition, have special survival value because they facilitate cooperation and communication between individuals within the group setting.

British anthropologist Robin Dunbar argues that many modules in man's enlarged neocortex evolved as a means to survive and thrive in large, complex social groupings. Some behaviors associated with living in large groups are reciprocal altruism, deception, and coalition formation. One can argue that man's development of language would also facilitate cooperation within large social groupings. In each case a cluster of neurons (a module) is dedicated to facilitate a particular behavior.

Unlike supercomputers, people with so-called high IQs are not capable of performing all tasks faster and better than individuals with lower IQs. Intelligence is not a one-dimensional quantity. A particular individual may be able to do complicated math problems better and faster than someone else, but may not be as capable with respect to language skills. Someone might be exceptionally musically talented, but may be lacking when it comes to spatial relationships. Indeed, some people can possess a specific talent or aptitude, like being good at playing chess or having an "ear" for language, and yet these same individuals need not display or possess the same expertise with various other endeavors—or even seem overly "bright."

A case in point would be "Williams Syndrome." Individuals with Williams Syndrome typically have low IQs—between 50 and 70. They may have trouble telling their right from their left and they can't perform simple addition. And yet people with Williams Syndrome often turn out to be gifted musicians and voracious readers. They are also fascinated by people in general and are remarkably empathetic.

On the other hand, individuals with autism do not necessarily have low IQs but they consistently find it difficult to grasp the rules of society or understand what other people are thinking and feeling. Because our brains are a collection of task-specific parts, human behavior is quite complicated, and assessing someone's intelligence is not a simple task.

Due to the modular nature of the brain, individuals vary greatly with regard to their particular strengths and abilities. This turns out to be very useful in terms of man's evolutionary success by ensuring diversity within a human grouping.

Neurological Mapping

Although there is a digital (binary) aspect to whether a neuron "fires" or not, for the most part, our brains function in an analogical fashion. Neurogical pathways are established and organized in a way that "correspond" to outside events. We might think of it as some sort of analogical "mapping" process—a neurological "mirroring" of the outside world.

In order for us to better understand the concept of neurological mapping, let us first consider an example of a non-neurological mapping—the Jacquard loom (invented in 1801). This automated loom used a chain of punch cards to direct the loom to weave the particular desired pattern on the fabric. The location and spacing of the holes in the punch cards permitted the hooks on the loom to raise or lower the warp thread, thus creating the desired pattern on the fabric. To look at the punch cards, you could never guess what the pattern on the fabric was going to look like, but the pattern of holes on the punch cards did translate eventually to the fabric's pattern, and as such, was a "mapping"—an analogical representation of the desired fabric pattern.

Another example of non-neurological analogical mapping has to do with AM and FM radio signals. (AM stands for amplitude modulation, and FM stands for frequency modulation.) The basic process is that the sound waves going into a microphone modify or modulate the electrical current in the microphone so that the variations in the electrical signal mimic (mirror) the variations in the sound waves. The modified electrical signal is sent to an antenna and broadcast through the air, eventually reaching our little transistor radios. Inside the radio, the variations in the radio signal modulate the physical vibrations in the radio's speaker, producing the sound waves that eventually reach our ears. During each phase, one type of signal (information) is made analogous to a different type of signal; first sound to electrical, and then electrical to sound. Information is not digitized (turned into 0's and 1's) at any point along the way. Instead, information is physically "mapped," "mimicked," or "mirrored" from one phenomenon to the next. This is an example of an analog process, not a digital process. However, many radio stations are in the process of switching to a digital signal.

Returning to our human brain, let us now consider a professional basketball player practicing foul shots over and over again. For starters, our basketball player tries to picture in his mind the path of the ball as it sails from his hand through the air to the hoop. When a practice shot does finally go in the basket, the brain tries to correlate the successful body, arm, and hand movements to the specific neurological signals that caused the correct flight path. With each successful shot, the brain modifies, rewires, and fine-tunes itself, so that the new neurological arrangements and connections are in some way analogous to—correlated to—the complicated physical movements that are required to shoot a basket.

Many athletes of various other sports, such as diving or skiing, go through a mental exercise of picturing the correct execution of their task in their mind's eye. In doing so, they are trying to correlate, or map, what is going on inside their brain with the desired implementation of their athletic endeavor.

Consider a somewhat more academic example, like learning how to read. Learning how to read does not involve converting the letters and words on the page into a series of 0's and 1's like a digital computer would do. Information is not digitized. It does instead involve the establishment of neurological pathways and connections in response to the visual input (letters and words) coming from the written page. These neurological modifications, these mappings, these correlations, are what's happening in our brain when we learn.

On a neurological level we are talking about two events being experienced or "recorded" on two distinct sets of neurons, followed by a neurological pathway being established between the two. We perhaps sense what has occurred in our brain and sometimes say that we have finally "made the connection." Learning has occurred!

Correlations, Patterns, and Causality

Imagine that somehow you find yourself on a very distant and strange planet, one that may not even be part of our known universe. It is your brain's job to help you cope and survive, so your brain immediately starts trying to figure out how things work on this strange planet. It does this by looking for associations between one event and the next.

Imagine, for example, that when you let go of a ball on this strange planet, it moves sideways (instead of down). Having seen this phenomenon several times, your brain would see a *pattern* and start to *correlate* the letting go of the ball with the sideways movement. Your brain might then surmise that there is some sort of invisible force on this strange planet that has indeed *caused* the ball to move sideways. (Of course, back on good old Earth the ball would have instead been pulled downward instead of sideways, and our brains would have come to the conclusion that on planet Earth there is indeed some invisible force—gravity—pulling the ball downward.)

Suppose also that on the aforementioned strange planet every time you hear a loud noise the temperature seems to drop. Your brain would begin to see a *pattern* and *correlate* the loud noise with the drop in temperature and might eventually assign *causality* to it.

Let us return to Earth and think about prehistoric man seeing dark clouds roll in and feeling the wind start to blow. Perhaps he would also see some lightning or hear thunder. Having experienced these phenomena several times before, he would begin to *correlate* the clouds, wind, lightning, and thunder with a dangerous storm that was approaching. He would have deduced a *pattern*, and presumably would have sought out appropriate shelter.

Or imagine prehistoric man striking certain kinds of stones together *causing* one of them to chip and flake revealing a razor sharp cutting edge. Or picture him throwing a rock at a dangerous carnivore *causing* it to run away from its kill, leaving behind some fresh meat for our prehistoric man and his family.

In general, our brain seeks to associate one event with another. Michael Shermer, in his book *The Believing Brain* (2012), refers to this as the brain's search for "patternicity." This attempt to recognize associations, correlations, patterns, and causality—amongst the myriad events surrounding us—is one of the most important and essential tasks of our brain's neocortex.

> *"The human brain is an incredible pattern-matching machine."*
> *- Jeff Bezos (founder of Amazon.com)*

Too Much Cause and Effect?

There is, however, much randomness and chaos in our world. And not every event or happening is the direct cause of some other event or happening. For example, when something unfortunate happens to you after a black cat crosses your path, it doesn't mean that black cats *cause* bad things to happen. Similarly, crossing your fingers does not necessarily *cause* good things to happen to you. Because our brain constantly seeks *causality*, humans have a natural propensity to assign blame. This can often lead to superstitions and the belief in "lucky charms." We are, to a large extent, a rational animal looking for causality in an incredibly complicated and often chaotic world. Our brains are hard-wired to seek causality, even when there is none.

Our Operating Platform

As previously stated, our brains do not function exactly like a sophisticated computer. But there is a similarity between the human brain and a computer that should be discussed. Every computer has an "operating system," or "operating platform," from which all operations and tasks are managed. The operating platform is the organizational framework that governs all the various functions and capabilities of that computer. As stated earlier, our brains try to figure out how the world works by looking for correlations and connections in the outside world. And like the computer, our brain needs an overriding framework to organize itself and make sense of what it observes and experiences. For humans this operating framework is what we might call a "worldview." This worldview might include things like how the world was created and all the various laws and theories of both the physical and the social sciences. It might also include some sort of a moral code or a canon of religious principles.

Worldview as a Narrative

The term "narrative" can be used to help describe this concept of a frame of reference for your brain. We can all perhaps agree that a narrative is some type of a story—written or spoken—that describes events, experiences, and the like. On one level, many stories embody a kind of social cause and effect in that they contain illustrations of social situations from which lessons are to be learned. Human beings love to read novels and watch TV shows about people and various social situations. We enjoy hearing stories and telling stories perhaps because the way that a narrative unfolds reflects the way life unfolds. And so evolution has seen to it that listening to and participating in narratives are neurologically rewarding activities.

On a grander scale, the concept of a "worldview narrative" is useful because it is a story or a description of reality itself—a story about how the Earth began, how humans got here, and how the world works. As previously mentioned, ancient man was surrounded by chaotic events such as earthquakes, volcanoes, thunder and lightning. He observed strange happenings such as the tides, seasonal variations, a waxing and waning moon, and a mysterious sun that seemed to revolve around the Earth, deserting him at night. Throughout all this wonder and chaos, the brain looks for correlation and causality, and seeks to create a narrative to explain it all.

The Hebrew bible is an example of a worldview narrative that seeks to explain the mysteries of the universe and provide meaning to life. An individual who subscribes to the Hebrew bible might have a worldview narrative that includes a belief in the creation story found in Genesis— that God created the world in six days about 6,000 years ago. He or she might believe that the rules for moral behavior are described by the "Ten Commandments." (Exodus 34:28 and Deuteronomy 10:4) This person might hold certain beliefs about the various natural and social sciences, have a set of political views, and possess opinions on the various issues and problems of the day.

From a practical, operational point of view, we can perhaps think of a worldview narrative as being a kind of story that helps direct and guide our actions—our brain's operating platform. Looking at the notion of a worldview narrative from a philosophical point of view it can be considered an effort to assign meaning and essence to man's existence.

This doesn't necessarily mean that each of us, having established a worldview narrative, necessarily thinks that we have it all figured out. However, any information or data that runs contrary to, and challenges our worldview, is often met with a great deal of caution and scrutiny. Ideally, all the various components that make up a person's worldview should compliment and validate each other. Each part should reinforce and serve to buttress the other parts. Each piece should add clarity and definition to the "big picture."

As a complete, integrated whole, our worldview narrative is supposed to make sense. Modifying ones worldview is therefore serious business. After all, this is the framework that guides us through life and from which many important decisions are made. If even only one particular aspect of our worldview is questioned, or seems to be in doubt, we might fear that the whole integrated package is in danger of unraveling.

On a neurological level, our brain floods itself with adrenaline and dopamine (positive feeling) when we think we are right. However, if our correctness is in doubt, our brain instead floods itself with cortisol (negative feeling). As a result of this cortisol, the executive functions that help us with advanced thought processes (like decision making, organizing information, and problem solving) are inhibited, and our amygdala (emotional and instinctive brain center—home of the "fight or flight" response) comes to the fore. It is therefore easy to understand why most people, when confronted with any new and potentially contradictory information, react very strongly to any challenge to their long held belief system—their worldview narrative. For this reason, many people seek out only information and opinions that conform to their worldview, especially when it comes to politics.

We can perhaps agree that in Paleolithic times, and in dangerous times in general, having our amygdala kick in (fight or flight) can be a very useful and practical thing. However, having our amygdala take charge when we are dealing with some very complicated situations—like the ones often encountered in modern times—can be counter-productive. Sometimes we jokingly say that someone is acting like a "caveman" or a "knuckle-dragger" when their thinking and behavior are more befitting prehistoric times than modern times.

Fact or Fantasy

Even though most people believe that their worldview narrative represents an accurate analysis of reality, it can in fact sometimes be somewhat fictitious. Most worldview narratives are not based on thorough research and analysis. Most people don't have the time or patience to do so. But everybody instinctually understands that they must formulate something. Man needs to put things in order—to make sense of the world.

Scientists sometimes say, "Nature hates a vacuum." For our brain, any worldview is better than none, even if it is in large part based on myth.

One might assume that the truer and more accurate one's worldview narrative is, the more successful one will be. That may be true in general—for most people, most of the time—but sometimes raw scientific accuracy is not strictly necessary for success.

For example, believing—contrary to all scientific evidence—that the Earth is only 6,000 years old (as suggested in Genesis) would not necessarily prevent an individual from becoming a successful clothing designer, musician, or business owner. And one can conceive of a situation where an individual's estimation of his or her talent is perhaps a bit inflated. But armed with a positive self image, with perseverance, and hard work success is possible—maybe even probable. Similarly, if you are a blonde and you believe blondes have more fun, this may have a beneficial effect on your socialization repertoire.

It would seem that there is an evolutionary benefit to having the capacity to imagine. Delusions and myths can sometimes be very comforting and quite useful. People do regularly interpret "facts" to fit their existing narrative. Even if an individual's worldview narrative is not totally accurate, it does provide a frame of reference from which to live one's life and get on with the business of survival.

Believing that the Earth is flat or believing that the stars control the future might tend to limit one's progress in today's modern world, but it wouldn't have been such a disadvantage in prehistoric times. And so it should not come as a big surprise that our Paleolithic mind, while capable of complex, rational thought, is also comfortable with misinformation and myth.

Conspiracy Theories and Myths

"The human brain has evolved the capacity to impose a narrative, complete with chronology and cause-and-effect logic, on whatever it encounters, no matter how apparently random." - Robin Marantz Henig (Writer for New York Times Magazine)

Man's need to assign causality can sometimes make him susceptible to conspiracy theories. Conspiracy thinking is a fairly common phenomenon. There is no shortage of conspiracy theories regarding President Kennedy's assassination or theories about Jews trying to control the world. Regardless of whether it is accurate or not, a conspiracy theory is an explanation for an occurrence. And as stated earlier, our brains do try to find causality in what we observe around us. You may have heard of the expression "everything happens for a reason." That's your brain looking for causality. Any explanation, even an inaccurate one, is often more comfortable and better than no explanation at all. The unknown, disorder, or chaos can be very unsettling to the human mind.

Human history and culture are filled with fables and myths. For the most part, fables and myths are not meant to be scientifically accurate. Many times these stories or narratives are meant to convey information about the world around us, or perhaps teach a moral lesson. However, in today's scientifically advanced society, believing in fables and myths can have its limitations.

Although, generally speaking, having a realistic worldview narrative is more conducive for survival and success, there are a lot of very financially successful people with major inaccuracies in their worldview narrative. For example, not believing in human-caused climate change is not a handicap for individuals employed in the fossil fuel industry. To the contrary, it can be advantageous and very convenient to say the least.

Genetically Encoded Curiosity

As stated previously, *The Paleo Perspective* looks at human behavior through the prism of evolution. Besides looking for cause and effect, what else has evolution programmed our brain to do? According to Dr. Maria Montessori, an important function of the human brain is its propensity to explore. (Dr. Montessori was the first woman in Italy to become a medical doctor, and was the founder of the "Montessori Movement" and Montessori schools.) Humans, as well as many other animals, are curious creatures. Being curious and having a proclivity to explore one's environment can be a very useful adaptation. It is not hard to envision how exploration can add to one's knowledge and understanding of one's surroundings, thereby giving the individual additional information to enhance the probability of success and survival.

"The important thing is not to stop questioning. Curiosity has its own reason for existing." - Albert Einstein (German-born theoretical physicist)

We sometimes observe animals, such as cats, exploring their surroundings. There is an old saying that "curiosity killed the cat." Yes, every once in a while a cat will get itself into trouble by going into dangerous places, but on the whole, a cat's inclination to explore is very beneficial to its survival. Similarly, it was advantageous for Paleolithic man to explore his surroundings. Exploration helped man discover new caves for shelter, new food sources, and helped him avoid dangerous situations.

Curiosity is encoded in our brain, and involves the release of pleasure-giving endorphins. It is not some sort of serendipitous bonus that makes our lives more interesting. It is an evolutionary adaptation, just like a chameleon's ability to change color.

An Inventor and Problem Solver

Another genetically programmed human behavior is our drive to solve problems through ingenuity and resourcefulness. Man has a strong inclination for technological inventiveness—a desire to constantly invent "a better mousetrap." Man is, without a doubt, the most creative creature on the planet. Putting aside for the moment his artistic creativity, which will be discussed in a later chapter, let us consider a tiny sample of his inventions through time.

The earliest and most important inventions were probably various stone tools and weapons. Stone tools found at Lake Turkana in Kenya are dated to be 3.3 million years old. Going forward with some other notable inventions: there was the wheel, the chariot, the windmill, rockets, the printing press, the telescope, the microscope, the refrigerator, the steam engine, the automobile, the cotton gin, the telephone, the computer, and the cell phone—to name just a few. Man is clearly programmed to be inventive—to "build a better mouse trap."

Our technological creativity is not a serendipitous fringe benefit to keep us amused. It is an evolutionary adaptation of great import. One could argue that some of our technological inventions, like sophisticated weaponry, now threaten our continued existence. But there is no disputing how evolutionarily beneficial technological creativity was to our species. There are other very clever and intelligent animals that share our planet. But when it comes to technology and engineering, man is clearly in a class by himself. Indeed, creativity and inventiveness were encoded in our DNA a long, long time ago.

You can't use up creativity. The more you use, the more you have."
- Maya Angelu (American writer, poet, and activist)

Theory of Mind

There is one more crucial and salient characteristic of the human brain that needs to be discussed, and that is what is called "theory of mind." There has been much research done on *theory of mind*. Among the developmental psychologists that have championed *theory of mind* are Josef Perner, Alison Gopnik, Henry Wellman, and Andrew Meltzoff. The basic idea of *theory of mind* is that humans can attribute mental states (thoughts, desires, and intentions) not only to themselves, but to other humans and other living creatures as well.

For example, we might think that someone thinks ill of us and is out to deceive us or do us some harm. Or perhaps we might think that another person admires us and thinks that we are very talented. In both examples we are imagining that other people have "mind," and that they are capable of having thoughts and feelings, just as we do.

Early hominins lived in large complex groups and acquired the ability to "read minds"—in a manner of speaking. By looking into the eyes of their fellow hominins they would theorize what they were thinking or feeling. They could read body language and reflect on the past actions of fellow group members. In the process, hominins began to develop better ways of making alliances and keeping track of one another.

> *"The face is the mirror of the mind, and eyes without speaking confess the secrets of the heart." - St Jerome (4th century Croatian theologian and historian)*

Man is a very socially complex animal. Living in large groups necessitates certain social skills such as the capacity for reciprocal altruism, deception, empathy, and coalition formation. Think of how difficult it would be to function in a highly social setting if humans had no interest or capacity to sense other people's thoughts, feelings, and motivations. Clearly, *theory of mind* would be a very advantageous evolutionary adaptation for socially complex prehistoric man.

It is interesting to note that according to a new international study (Christopher Krupinye and Fumihiro Kano, *Science, 2016),* great apes (chimpanzees, bonobos, gorillas, and orangutans) also demonstrate this ability. The authors of the study also state that humans develop this ability by age 4 or 5, and that *theory of mind* is central to a lot of sophisticated and social human behaviors.

Non-Humans

Man's application of *theory of mind* applies to non-humans as well. Humans are quite capable of assigning thoughts and feelings to their pets or even wild animals. They might think their dog is ashamed, bored, anxious, sad, or even depressed. In the case of a wild animal, we might assume that a snake has the intention to do us harm. Assigning "mind" to non-human creatures not only gives man the opportunity to anticipate danger, but also the possibility to manipulate a situation to his advantage. This is not to say that we are always correct in our assumptions about other creatures' thoughts and intent, but the very fact that humans theorize about intent gives us an important edge. This is somewhat like strategizing in a chess game by trying to anticipate your opponent's next move. You are not always right, but most of the time it is to your benefit to anticipate and plan ahead.

Inanimate Objects

One might wonder if this *theory of mind* phenomenon can be applied to inanimate objects as well. Do humans sometimes assign mind, intent, or "agency" to such things as cars, clothing, computers, stuffed animals, a flashlight, etc?

We sometimes think that our car has it in for us because it gives us a problem at the least opportune time. Think about that flashlight that flickers just when you really need to use it. Doesn't it seem like it has a mind of its own? Or perhaps you feel that your car has a benevolent personality, like a trustworthy, reliable friend that will never let you down. Think about how a child can assign a loving personality to a stuffed animal, with nothing but positive regard for its owner.

Throughout history, humans have assigned "mind" to inanimate objects such as the sun, the moon, a mountain, or the stars. Indeed, sun or moon worship was not uncommon in ancient cultures.

Invisible Entities

Theory of mind can also be applied to things that are totally invisible and beyond the scope of physical science. Humans can believe in spirits—both good and bad. Many people believe that their deceased relatives exist in some other world or dimension—a kind of spirit world—and are still capable of thoughts, feelings and intent. That is to say, they believe that their long-lost loved ones still have a mind. Belief in an afterlife can involve the concept of a soul—that part of a person that exists long after one's body has perished. It is hard to conceptualize a person's soul without at the same time thinking about the departed person's personality and mind.

For those who are inclined, the spirit world can contain other spirits besides one's relatives. Some people believe in the existence of an evil spirit—a devil. On a more positive side, according to a 2014 Pew Research Survey, about 89% of respondents believe in a benevolent god or some sort of universal spirit. Most people think of their god as being capable of thoughts, having a sense of fairness, and a notion of right and wrong.

Many people's concept of a god not only includes the notion of a thinking god, but it also includes the notion that this super being is omnipotent and in control of every cosmic event—big and small. One might be familiar with the often-used and commonplace expression, "God willing," used in all sorts of everyday conversation. The use of this expression is suggestive of man's belief in a world under the influence of some kind of invisible, omnipresent, and mindful supernatural being.

This brings us to the concept of "divination." Divination (from the Latin divinare, which means "to foresee, to be inspired by god") refers to a practice whereby the participants expect their god or surrounding spirits to influence the outcome of some (arguably) random events. A divination ritual might involve the "reading" of the pattern of chicken bones thrown on the ground, the cracks in heated turtle shells, the shapes of animal organs, or in more modern times the reading of Tarot cards. The theory behind these practices is that supernatural beings or forces will make their will known through these ostensibly random happenings.

A seemingly logical extension of this kind of thinking is the idea that the supernatural being—the god—might punish people for bad behavior or reward people for good behavior. For instance, the god might inflict mankind with a disease—such as AIDS—for immoral behavior. Or the god might reward his people with a military victory for righteous behavior, such as the victory of David over Goliath (Hebrew bible).

Santa Claus

This human capacity to imagine supernatural entities (with minds) is especially prevalent in youngsters. Children can have imaginary friends. Most children—below a certain age—believe in characters such as Santa Claus, the Easter Bunny, and other fantasy figures. Children are encouraged to behave properly because "Santa is watching."

Recently, something called "The Elf on a Shelf" has become popular. Based on a tradition Carol Aebersold began with her family in the 1970s, a cleverly rhymed children's book explains that Santa knows who is naughty and who is nice because he sends a scout elf to every home. During the holiday season, the elf watches children by day and reports to Santa each night. When children awake, the elf has returned from the North Pole and can be found hiding in a different location. In the child's mind, this little toy elf (sitting on a shelf in their bedroom or playroom) has a mind, which it uses to assess and report behavior to Santa. The stronger a child's belief in Santa, or his elf accomplice, the more effective this illusion is at encouraging good behavior.

The human brain does have a tremendous capacity to imagine things. This capacity allows us to plan things out ahead of time to help us manage both the physical and social world around us. We can imagine how to perform a physical task, such as making a tool. We can imagine a sequence of social interactions and plan accordingly. However, our ability to imagine does have some drawbacks. Our capacity to believe in myths can, at times, be counter-productive.

We will have a more complete discussion on religion per se in a later chapter. Suffice it to say that our belief in a god and a devil (whether or not they exist), Santa Claus, or the spirits of our deceased relatives is a function and extension of *theory of mind*.

For those whose concept of a supernatural presence is less anthropomorphic than Michelangelo's image of a white bearded God, or Clement Moore's pleasantly plump version of St. Nick, there is also the concept of a "cosmic mind." *Cosmic mind* is when we assign "mind," or some sort of mindful spirituality to the cosmos itself. An example of this sort of thinking can be found in the popular movie series "Star Wars" (1977). I refer you to the famous catchphrase "may the force be with you." In any case, assigning consciousness and willfulness to any supernatural anthropomorphic entity, or the invisible cosmic surroundings, are examples of *theory of mind*.

An Ancient Mind and Body

A *paleo perspective* on the human brain is that it is a 200,000-year-old design. Evolutionary biologist Gordon H. Orians makes note of this fact in his book *Snakes, Sunrises, and Shakespeare: How Evolution Shapes our Loves and Fears* (2014). He offers several illustrations of our ancient brain at work.

As a first case in point, Orians points out that studies show that people who have never encountered a snake still report an intense fear of them— far greater than their fear of guns, speeding cars, or other modern dangers that in fact pose a more serious risk. Similarly, in memory experiments, people are better at recalling animal tracks than other unfamiliar objects.

He also believes that certain landscapes and other natural features, like sunsets, appeal to us because of the role they once played in keeping us safe. Nighttime was a time of extreme danger to prehistoric man. And so sunsets served as a warning about the approaching danger and therefore needed to be a real attention grabber. These are what we might consider residues from our evolutionary past.

The concept of a prehistoric brain is similar to the notion that our bodies also have some prehistoric residues—sometimes referred to as vestigial remains. Consider the following:

- Our ears have several features—Darwin's Point and auricular muscle— that are remnants from a time when our ears had the ability to turn and move about like they do in more primitive species.
- Our coccyx—3 or 4 small, fused bones at the bottom of our spine—is what is left of a tailbone.
- Our appendix—virtually useless in modern times—may have once aided our primate ancestors with the digestion of their cellulose-rich plant diet.
- Erector pili—better known as "goose bumps"—are tiny muscles in mammals that cause the hairs to stand straight up. Humans are relatively hairless, but for furry mammals this serves two purposes. It helps to retain heat, and also serves to make the mammal appear bigger during confrontations.
- The "grasp reflex" is prevalent in most modern human infants. This vestigial remain was very useful in our evolutionary past, and is common amongst primates where the young need to hold onto their mother's fur as she moves about.

And so, both our bodies and minds have many residues of times long gone. Accepting this reality helps us understand human nature better, and helps us deal with the many complexities of modern life in a more informed way. Such is a *paleo perspective.*

"The brain is conceded to be the master organ of the body, the regulator of life, the source of human progress." – Sir Frederick Tilney (15th century Lord of Lincolnshire, England)

PART II

SURVIVAL

Competition or Cooperation?

CHAPTER 9

SURVIVAL OF THE INDIVIDUAL

"The instinct to survive is human nature itself, and every aspect of our personalities derives from it. Anything that conflicts with the survival instinct acts sooner or later to eliminate the individual and thereby fails to show up in future generations." - Robert A. Heinlein (American science fiction writer)

The individual's instinct to survive is primary. As individuals we are constantly seeking to enhance our chances of survival. If the individual doesn't survive, the opportunity to reproduce is jeopardized, and survival of the species is threatened. *The Paleo Perspective* takes the position that the instinct to live is imbedded in every cell of our being. It is encoded in every strand of our DNA. The instinct to survive is of course not unique to our species. If we examine our evolutionary family tree we see that the instinct to survive was present in the various ancestral species that predate Homo sapiens.

Having said that, there are circumstances that involve mental illness, extreme religious behavior, or situations where living is so painful and void of any hope that people choose to end their lives. The Center for Disease Control (CDC) reported that in 2010 the suicide rate in the United States was about 12 per 100,000 people. That equates to about .01%. Looking at this statistic in a positive way, 99.99% of us chose to continue living. Even one suicide is unfortunate, but as you can see, the desire to live is extremely high. Also, bear in mind that the suicide rate is generally much higher for people over 85 than it is for people in their twenties. Ancient man for the most part did not live much past thirty years old anyway, so one wouldn't think that suicide was a major concern for our ancient ancestors.

"Who wants to die? Everything struggles to live. Look at that tree growing up there out of that grating. It gets no sun, and water only when it rains. It's growing out of sour earth. And it's strong because its hard struggle to live is making it strong. My children will be strong that way." – Betty Smith (Author of *A Tree Grows in Brooklyn*)

We have all heard of or know of examples of individuals who persevere with a devastatingly horrible disease, or people who survive a traumatic accident.

A case in point is an article that appeared in *Newsday* (March 22, 2014). The article concerns a woman named Charla Nash, who had been horribly disfigured in a 2009 chimpanzee attack. The chimp bit off both her hands, and bit her face so severely that she is currently blind and has had to have a face transplant. Charla's hand transplants were unsuccessful and she continues to struggle with complications resulting from her face transplant. Ms. Nash did not give up on life. She sued the state of New York so she could pay her medical bills and "have a chance to live a comfortable life."

There is also the story of Aron Ralston who was hiking in Utah in April of 2003 when a dislodged boulder pinned his right hand against the wall of a narrow canyon. He tried for days to free his hand to no avail. Out of food and water he finally, courageously, managed to break his radius and ulna bones, and then proceeded to sever the lower part of his forearm and right hand with a two-inch knife. During this ordeal Aron lost 40 pounds, was forced to drink his own urine, and lost 25% of his blood supply because of the amputation. But he managed to free himself and hike back to safety.

Finally, let us think about the Jews who survived the horrific conditions in Nazi concentration camps during World War II. Many were beaten and starved to death—but many persevered. No one could survive that kind of hell without having a strong instinct and will to survive. There are countless other examples of individuals facing extremely dire circumstances and choosing to live—choosing to survive.

Making a Living

As we continue our discussion on surviving, let us turn to the very practical subject of how people make a living. It is perhaps telling that we use phrases like "making a living," "earning a living," or "livelihood" when discussing this topic. Indeed, it is no accident that these phrases contain the word "life" or "live" in them. Along the same lines, an individual is often referred to as the "breadwinner," and his or her job as the "meal ticket." Money itself is sometimes referred to as "bread." This should come as no surprise because having a job puts food on the table—and everybody needs food to survive. Everybody understands how essential to their survival their job really is. Anyone who has lost his or her job for any length of time knows how frightening that experience can be.

And so it shouldn't come as a surprise that people will do just about anything to earn a living—to survive. They will lie, cheat, and steal. They will become drug dealers, pimps, prostitutes, pirates, or human traffickers. Anybody who reads the daily newspapers is aware of the incredibly despicable things people do to ensure their survival. We read about people stealing copper wire, cheating Medicare, selling pornographic pictures on the Internet, creating illegal Ponzi schemes, making counterfeit money, extorting, blackmailing, and even contract killing. The list goes on and on. Given man's instinct to survive, we shouldn't be surprised that crime is higher in poor neighborhoods.

Even if an individual has a perfectly respectable, legal job, there may be some lying, cheating, or stealing that goes on at that "respectable" job. Some people might cut corners just a little, and some may bend the rules a lot. No matter what rules or regulations society puts into place, it is human nature to consider avoiding or circumventing those rules to enhance one's individual well-being—to enhance your survival prospects. Humans are capable of doing anything to survive.

You might have heard the phrase "everyone has his price." That is to say that if you are desperate enough, there is no telling what you are capable of doing. We would all like to think that our moral values would prevent us from doing something really terrible, but nobody really knows how they would react if the situation became desperate enough.

The following is an excerpt from an article appearing in *The New York Times* on March 29, 2014:

> *Dakope, Bangladesh - When a powerful storm destroyed her riverside home in 2009, Jahanara Khatun lost more than the modest roof over her head. In the aftermath, her husband died and she became so destitute that she sold her son and daughter into bonded servitude. What is more, Ms. Khatun now lives in a bamboo shack that sits below sea level about 50 yards from a sagging berm. She spends her days collecting cow dung for fuel and struggling to grow vegetables in soil poisoned by salt water.*

The point is that most people will do whatever it takes to sustain themselves. Man is genetically programmed to survive. However, by saying that man is genetically programmed to survive, we are not excusing or condoning any illegal or depraved act. We are simply stating the obvious. Homo sapiens, like other species, are the result of millions of years of evolutionary modifications, and that the instinct to survive for any species is absolutely essential for success. A *paleo perspective* encourages us to accept this aspect of human nature as crucial. Ignoring this aspect of human nature can lead to disappointment and frustration.

For instance, extreme poverty will lead to all sorts of crime. As stated before, people will do anything to eat—to survive. Looking at crime strictly as a moral failing ignores the most fundamental, underlying cause of crime. We are not saying that there isn't a moral dimension to crime. What we are saying is that we shouldn't be surprised when people don't behave in a legal or moral fashion when their very survival is threatened. People will do whatever they have to, regardless of whether it is moral or legal, to survive.

Those Survival Genes are Always Working

It may seem that we have painted a rather black and white picture. One does whatever it takes—lie, cheat, or steal because it's a question of life or death. What about all those everyday situations where your life doesn't hang in the balance? What about the guy who has a steady job, but steals some materials from the job site, or helps himself to some office supplies? Surely he doesn't think that his very survival depends on his minor thievery. After all, he has a source of income—he is "making a living." Perhaps he is just being greedy.

Employing a *paleo perspective* requires us to focus on man's genetic makeup—the same genetic makeup of our ancient ancestors. Homo sapiens were designed for a time when much of what they did was indeed a matter of life and death. Our instinct for survival had to be strong. If it wasn't, we wouldn't be around 200,000 years later.

Even though most everyday situations aren't life or death situations today, your "survival genes" are still hard at work, lurking in your subconscious mind. And so all throughout our modern lives, we still feel the urge to do whatever it takes to enhance our prospects for survival. Let us look at some routine things we do each day, and look for those behind the scenes genetic forces at work.

During a typical day most people start off with some kind of breakfast, maybe take some vitamins, think about the weather and decide what is appropriate to wear. Most of us obey the traffic rules, try to get to work on time, and do our best while at work. We might look for a little overtime, join a union, sign up for health care coverage, or maybe even suck up to the boss a little. Perhaps you stop by the bank on your way home to put a little money away just in case a rainy-day emergency presents itself. After work you might "hit the gym" or attend some evening college classes. When the day is done you might try to get a good night's sleep to help prepare for the following day.

When you think about it, each of these activities enhances your chances to survive in some small, measurable way. Indeed, most of the things we do hour-by-hour, day-by-day, week-by-week, and year-by-year involve behaviors directed towards enhancing our survivability in some way, shape, or form. We are all coded to survive. This is all part of being human.

But, How Much is Enough?

Do squirrels ever stop gathering nuts? Do they ever come to the conclusion that they have enough nuts to make it through the winter and stop hoarding? A brief search on the Internet seems to indicate that squirrels will hide as many nuts as they can, just to be on the safe side. Having more nuts increases their survival chances.

In a modern market economy, making more money is a reflection of our instinct to survive. One might wonder how much money, how many cars, houses, and other various material things are necessary or sufficient to satiate man's instinctual urge to survive? After making the first million dollars, why do most people endeavor to make the second million—or more? It is rational to want a roof over one's head, but do we really need a thirty-room mansion and several vacation homes? Most people in the developed world need some form of transportation to go about earning a living, but does that form of transportation have to be a Rolls Royce or a Mazerati?

"Greed is a basic part of animal nature. Being against it is like being against breathing or eating. It means nothing."- Ben Stein (American writer, lawyer, actor, and speechwriter)

The Paleo Perspective reminds us that our genetically inspired survival impulses are always with us, down deep in our psyche. More does feel better. More does feel safer. The drive for bigger and better, for more and more, is a natural urge to build a reservoir of safety in our quest to survive. Two million in the bank is more of a safety net than one million. Perhaps we can liken this human urge for "more" to an animal building a big reservoir of fat for the approaching winter. More fat enhances survivability. Human desire to accumulate wealth is similar to nut hoarding and fat storing. Wealth enhances survivability. Self-preservation is at the core of much of what we do.

Contradictory Impulses

Not everything human beings do is positive and adds to their survival prospects. At times Homo sapiens engage in behaviors and activities that are not only detrimental, but are sometimes life-threatening. Man is far from perfect.

People smoke cigarettes, take illegal drugs, drink too much alcohol, over-eat, under-eat, gamble too much, drive too fast, and engage in lots of other risky behaviors. Some bad habits involve addictions, some involve some type of mental illness, and some involve various distortions of logic to rationalize destructive behavior. Sometimes these counterproductive actions get the best of us, like overdosing on a drug, or getting tobacco induced lung cancer.

But even when engaging in these assorted kinds of destructive habits, people can simultaneously be doing other things that help them stay alive. An individual might like to drive too fast, but might also visit the gym religiously. An individual might smoke cigarettes, but is very conscientious and a real go-getter at his or her job. The instinct to survive is still there, below the surface, but it is sometimes competing with numerous other negative propensities and forces.

"I'm a beautiful mess of contradiction, a chaotic display of imperfection."
- Sai Marie Johnson (American author and poet)

The American Social Security system is a case in point regarding this particular aspect of human nature. Everybody understands that one should start saving money for retirement as soon as possible, and that putting a little money aside each pay period will enhance their future quality of life and long term survivability. But many people, even with the best of intentions, find this weekly exercise in delayed gratification difficult. Understanding and accepting this aspect of human nature, the government steps in and "forces" people to contribute to a social security trust fund (social security tax), ensuring that everyone will have some minimal level of income for their retirement. This human shortcoming is alluded to in the New Testament (Christian bible): "The spirit is willing, but the flesh is weak." (Mathew 26:40).

There are, of course, various moral and social dimensions to our behaviors. Religions have guidelines and societies have their rules as well. Human behavior is perplexing. There are many internal forces at play at any given time, and predicting what someone will do in any given circumstance is often extremely difficult and quite frustrating.

> *"I survived because the fire inside me burned brighter than the fire around me."- Joshua Graham (American author, fiction)*

CHAPTER 10

SURVIVAL OF THE GROUP

"Snowflakes are one of nature's most fragile things, but just look what they can do when they stick together." –
Vesta M. Kelly (American compiler of quotations)

Man is a social animal, and by far the most socially complex animal on the planet. We are all connected in some way to each other and to all other living things. Our brains are wired to be social because in many ways our very survival depends on our social skills. In the previous chapter we talked about man's instinct to survive as an individual. A *paleo perspective* recognizes the fact that man's evolutionary success is not only based on his strong individual will to live, but also on his intrinsic understanding that his own life depends on the survival of his fellow group members.

In an earlier chapter we discussed how essential man's Paleolithic hunter-gatherer groupings were to his success. Early man hunted together, foraged for food together, sheltered together, raised children together, and fought together to ensure the group's continued success.

It is important to emphasize here that the evolutionary forces of natural selection have acted not only on the individual, but also on the group. For highly social animals, such as man, group well-being is just as important as individual well-being—maybe more so.

Man the Socialist

Although some might object to the use of the term, man is a "socialist" by design. Consider all the various social groupings people belong to. We are members of families, clans, tribes, towns, states, provinces, and countries. We may be members of the North Atlantic Treaty Organization (NATO), the European Union (EU), or the United Nations (UN). Human groupings are quite ubiquitous, but so varied that we find it necessary to employ many different words to capture their subtle differences. Assembly, band, cadre, caucus, clique, congregation, collective, crew, crowd, fraternity, gang, klatch, mob, platoon, sect, sorority, team, and troupe are just a few—each with its own nuance and inference.

Some of man's group associations are automatic. You don't get to choose who your natural parents are, and so by default you are a member of a nuclear family. By virtue of living in New York State you are automatically a New Yorker. You can choose not to be a New Yorker by moving to a different state. But then you might be considered, for example, a Texan or a Californian. Unless you are stranded on some remote island, it is almost impossible not to be a member of one group or another.

However, man does freely choose to be part of various groups. On a local level, people join social service organizations such as the Lions, Rotary, or Kiwanis clubs. We join hobby or sports clubs dedicated to stamp collecting, photography, books, bowling, golf, or bicycling. As workers we are members of professional organizations and trade unions. We might be a member of the Army, Navy, or Air Force. And our religiosity can make us members of a church, a mosque, or a synagogue. The list can go on and on. One might even suggest that man is continually in search of a "tribe" to belong to. The point is that man freely forms associations with his fellow human beings, associations that meet both his personal interests and his survival needs.

Man the Confederate

Man is a "pack" animal—a confederate by nature— and does not need a particularly strong incentive to identify with a group. (The use of the term "confederate" refers to being united with others in a common cause.)

Back in the early 1970's, a British social psychologist named Henri Tajfel did a series of studies on intergroup discrimination. Individuals were randomly separated into groups using the toss of a coin or some similar, arbitrary method. Tajfel found that even though group formation was open and totally arbitrary, people favored their own group and discriminated against the other group. Man seems to have this inclination to want to be part of a team, even if being a member of said team is quite whimsical.

We might also say that man is quite "tribal" in the sense that he very quickly and naturally forms or identifies with a particular group. For instance, one might self-identify as a fan of a particular sports team. Most people who live in the New York metropolitan area are a Yankee or Met, Giants or Jets, Rangers or Islanders fan. People do expect you to have a preference. If you do not follow sports you are expected to be involved with something else (for example being a Star Trek groupie). There is an expectation that you self-identify with some grouping or another. You are in the minority if you don't.

We form "teams" quickly and put on various "uniforms" such as a business suit or "hippie" beads and headband. And we use symbols to declare allegiance to our team. Perhaps it's a peace symbol, or the symbol for the NRA. By displaying these symbols we are declaring our loyalty to the group. We are saying that we think in a particular way, and are to be grouped accordingly. (Birds of a feather flock together?)

Street gang formation would be another example of man's natural tendency to group associate. Gangs employ colors and symbols as a means of group identification. At the writing of this book many young adults are joining Islamic jihadist groups. There is evidence that their motivation isn't strictly religious, but instead they are responding to a human urge to be part of something bigger than themselves—the urge to be part of a group. Indeed, being part of a group can trigger the release of dopamine (a chemical in the brain), making it a neurologically rewarding experience. Group identification helps define who we are, and can be part of our worldview narrative.

So strong is our instinct to group that sometimes people align themselves with groups that seem detrimental to their own welfare. An example of this is the fact that several thousand African-Americans were reported to have fought for the Confederacy during the American Civil

War (John Stoufer, *Theroot.com*, 7/2015). Some slaves cooperated with the Confederacy at the point of a gun. Some did so as a matter of self-preservation when under attack or to better their status in a segregated South. But they also group-identified with their plantation, their State, and the South. To this point, *The New York Times* (July 20, 2015) had a picture of a black man named Anthony Hervey—in a Confederate uniform—waving a Confederate battle flag in an effort to honor black soldiers who served the South during the Civil War.

The Pain of Loneliness

Not only do humans desire to be part of a group, but they also go out of their way not to be alone or even feel alone. They talk to their pet dogs, cats, and birds. They talk to their plants and even inanimate objects. In the movie *Cast Away* (2000), actor Tom Hanks is marooned on a deserted island. Eventually he gets quite lonely and finds it helpful to talk to a volleyball that he names "Wilson."

Nowadays, if there are no other living things around, people will turn on the TV or listen to the radio. We see people with headphones, with a smart phone, with a tablet or laptop. Homo sapiens want to stay connected with other humans any way they can.

In fact, feeling alone is downright painful. Some people commit suicide because they feel utterly alone. If we want to punish someone, who may be already separated from the rest of humanity because of their incarceration, we put them in solitary confinement. Besides the death penalty and physical torture, solitary confinement is one of the most severe forms of punishment. Many people do consider it a form of torture. People have been known to go insane if left totally alone for long periods of time.

"No man is an Island." – Thomas Merton
(American Catholic writer, theologian, and mystic)

Let us agree that man is a highly social creature who hates to feel alone. But we are not alone on some island, and the Earth is a fairly populous place. The United Nations estimates that as of August of 2017, the planet had 7.5 billion inhabitants. Some are spread out in rural settings, but many people find themselves in urban settings. For example, about 80% of the people living in the United States find themselves living in some rather densely populated cities and urban areas.

Living in densely populated groupings does have some inherent advantages, however, it does require a good deal of cooperation. Let us look at some of the benefits of the group format and the necessary cooperation required for group success.

Two Heads are Better Than One

There are many potential benefits to operating as a group. Perhaps the most obvious one is that because it has multiple members, the group has access to the various talents that exist within the group's constituency, and thereby minimizes individual shortcomings.

For example, not all the males within a grouping need to be great hunters. There just needs to be a few good hunters within the group to ensure successful hunts. Not all individuals need to be great fighters or have a great sense of direction—just a few. Not all the females need to be great mothers or great cooks. There just needs to be enough females with the right instincts and the right skills to help out and show less experienced females how it's done. Intellectually, not everybody has to be a genius, there just has to be enough smarts within the group to figure things out. Not everybody needs to be a risk-taking explorer. All you need within the group is a few to take those risks and discover new territories and new hunting grounds. Groups, as a whole, benefit from the assorted talents of its various members. In diversity there is strength.

"None of Us Is As Smart As All of Us." - Ken Blanchard
(American author and management expert)

Furthermore, being part of a group means that members can brainstorm, share ideas, learn from one another, and spur creativity. They can support one another, motivate one another, and can learn to trust one another. Problem solving is almost always enhanced by group cooperation. And not all of the members need to be cooperative, so as long as there are enough cooperators within the group to keep the uncooperative members in line.

Wisdom of the Crowd

The sheer number of members within a group also offers a kind of social inertia, which serves to mitigate individual eccentricities. Extreme behaviors on the part of individuals can at times be detrimental to group success. Peer pressure and "groupthink" tends to rein in deleterious and undesirable conduct. We can perhaps think of this as a kind of group prudence.

Ultimately, we can think of the entire human race as one large grouping. The term "mankind" is often used to refer to the entire collective of humanity. In doing so, we are acknowledging the idea that our species can be thought of as a single organism, a super organism, which in and of itself endeavors to survive.

Having said all that, we must recall what was discussed in the previous chapter, that human beings also possess a strong instinct for self-preservation. The fact that evolution exerts pressure both at the individual level and at the group level is critical and helps explain many of the puzzling contradictions inherent in human nature. In an upcoming chapter on morality and altruism, there will be more discussion on how this evolutionary complexity results in conflicting urges within each of us. Let us now look at several illustrations of evolutionarily inspired group cooperation.

Order and Cooperation

Imagine you are living in the New York City area and you are going to a Met baseball game by car. You drive on the right hand side of the road (hopefully). You stop at stop signs, red lights, and for pedestrians in a crosswalk. You observe the posted speed limit, signal when turning, and in general do your best to comply with all the many New York State motor vehicle regulations. You are in fact being cooperative—cooperative with other drivers. Just imagine what chaos and injuries would result if everybody did whatever they wanted to while operating a motor vehicle.

When you arrive at Citi Field to see the Met game you would continue to cooperate in all sorts of ways. You would line up to purchase your ticket, sit in your assigned seat, and refrain from running out on the field during the game to greet your favorite player or to catch a ground ball. There might be twenty or thirty thousand fans at this game. Could you imagine what it would be like if everybody did whatever they wanted? Once again, it would be utter chaos and pandemonium.

Think about all the rules necessary for a school to function. You must get to class on time, raise your hand if you wish to ask a question, and dress appropriately. Or consider your place of employment. You have to get to work on time, punch out on time, return on schedule from coffee and lunch breaks, and obey all the many rules and regulations regarding the proper execution of your work-related duties. Businesses and corporations are good examples of group cooperation, coordination and collaboration.

People are cooperative in so many ways during their daily lives. We cover our mouth when we cough or sneeze and wash our hands after using the bathroom, all in an effort to prevent the spread of germs unnecessarily. By and large, we are conformists when it comes to clothing styles, hairstyles, music trends, and various other social trends. This is not to say that everybody follows these social conventions or that any of us do the right thing 100% of the time. But the vast majority of people cooperate with one another and with society at large most of the time in an effort to prevent chaos and unnecessary turmoil. It is our evolutionary instincts at work—subtly, unconsciously, promoting, and facilitating group success.

"Through the evolutionary process, those who are able to engage in social cooperation of various sorts do better in survival and production."
- Robert Nozick (American philosopher)

Teamwork

As previously mentioned, man has a Paleolithic urge to be part of a group—part of a "team." And when we use the term "teamwork" we are referring to the cooperation and collaboration between members of a "team" as they endeavor to accomplish a mutual goal. Teamwork happens on the athletic field, in a household, and on the floor of both large and small businesses and corporations. When even just two people cooperate, we sometimes say, "we make a good team." The sport teams and businesses that operate and function in the most cooperative manner are usually the most successful.

"Five guys on the court working together can achieve more than five talented individuals who come and go as individuals." - Kareem Abdul Jabbar (Retired American professional basketball player)

The Paleo Perspective on this matter is that Homo sapiens are fundamentally highly social and cooperative animals. Our ancient ancestors lived in cooperative groups for millions of years, and it is how Homo sapiens have survived for the last 200,000 years. It is one of our "signature" attributes. Perhaps we should call ourselves "Homo confoederata."

Whether it is at the town, city, national, or international level, our continued success as a species depends on our ability to cooperate with our fellow human beings. Indeed, our very survival depends on it.

"United we Stand, Divided We Fall"
- Aesop (Author of a collection of Greek fables, 500 B.C.E.)

CHAPTER 11

THE FORCES THAT UNITE

"Coming together is a beginning. Keeping together is progress. Working together is success."- Henry Ford (American industrialist)

As previously stated, evolutionary forces on our ancestors acted not only on the individual, but also on the group. Groups in which there was much cooperation, and groups that were strongly bonded together, tended to be more successful. Much of our success as a species is due to our proclivity to operate in cooperative groups. Having explored some of the evidence of early human cooperation and socialization, and having discussed man's inclination to be a member of a group, we can now turn to some of the phenomena that help keep human groups strongly bonded together. There are many aspects of human behavior that have helped human groups not only to survive, but also to thrive and flourish.

It turns out that there are many "glues" that help keep human groups bonded together. Let us focus on some of the more intrinsic or innate factors that contribute to human group solidarity. Let us consider those things that are part of our inherited human makeup and underlie man's exceptional capacity for complex group socialization and cooperation.

Kinship

Perhaps the most obvious example of human group association is that of the family, clan, or tribe. We are born into a nuclear and extended family unit, and we share genetic similarities with our bloodline. Tribal groupings and affinities exist today. Examples of modern day tribal identities include Gypsies, Jews, Native Americans, and many Arabic clans and tribes.

It is easy to understand one's affinity and fealty towards one's blood relatives. Clan loyalty was especially important in the past. Surviving in prehistoric times was difficult and tenuous to say the least. Early human survival was enhanced by man's cooperative and collaborative instincts toward his family troupe, whether it was group sheltering, group hunting, or caring for the more vulnerable members of the family, clan, or tribe.

This benevolent instinct toward genetically related individuals is illustrated when we consider the treatment of stepchildren as compared to the treatment of children by their genetically related parents. Neuroscientists Martin Daly and Margo Wilson, of McMaster University (Hamilton, Canada), analyzed data from the American Humane Association (an archive of child abuse reports in the United States). These records led Wilson and Daly to conclude that "a child under three years of age who lived with one genetic parent and one stepparent in the United States in 1976 was about seven times more likely to become a validated child-abuse case in the records than one who dealt with two genetic parents." Their overall findings demonstrate that children residing with stepparents have a higher risk of abuse, even when other factors are considered. Other studies (Crawford, 2008) suggest stepparents invest less in education, play with stepchildren less, and take stepchildren to the doctor less often. This phenomenon is sometimes referred to as the "Cinderella effect," and clearly illustrates our natural inclination to look more favorably on those who we are more closely related to.

Proximity and the Shared Experience

Similar to the bonding power of genetic kinship is that of "proximity" and shared experience. Living together, working together, facing hardships and danger together all enhance solidarity. For instance, being in an army platoon or a prison camp during wartime, is for many, a life-altering experience. People who have had such experiences sometimes refer to their cohorts as "brothers." Individuals who have joined together for a political movement or some other cause may think of their cohorts as "comrades." Even people who simply work together (in peacetime) sometimes refer to their co-workers as "family." For the sake of greater group cohesion, evolution has seen to it that prolonged proximity and shared experiences can indeed be powerful bonding forces.

Communication, Communication, Communication

Besides shared genes, man is hardwired to communicate. His multifaceted ability to communicate is critically important for human group cohesion. Insects and other animals communicate too. Spiders, termites and caterpillars communicate by using vibrations. Ants use their sense of touch by feeling the ant in front of them with their antennae. Silkworm moths use their sense of smell by releasing pheromones to attract a mate. Honeybees use a dance to communicate the location of a food source. Wolves howl, and bears leave scratch marks on trees. However, no species has quite the extensive repertoire of communication techniques that man has.

Before we explore all the ways that humans communicate, let us digress temporarily into the fictitious world of the *Star Trek* series. In that series there was a fictitious alien species called the "Borg." The Borg were an organic-cybernetic combination in which all the Borg individuals functioned as part of a singular organism which was called the "Collective." What enabled all of the Borg individuals to function as a singularity was the ability of each individual to communicate telepathically with all other Borg members instantly, simultaneously, and continuously—what you might call a "shared consciousness." They could literally read each other's minds, and thus their level of intra-species communication was so perfect that group cooperation and collaboration was absolute. With such group unity and solidarity, the Borg were indeed a very tough adversary for the all-too-human crew of the Voyager and their captain, Kathryn Janeway.

Humans of course cannot communicate with each other telepathically. But over the past several million years we have learned to communicate our thoughts and emotions in quite a number ways. It should be emphasized once again that in order for any social species to function smoothly, there must be effective ways for them to communicate. Socially organized animals such as ants, bees, African wild dogs, and mole rats have all developed highly efficient ways to communicate. Man is no different. For human groups to function optimally, group members need to know what is going on in each other's mind as much as is possible.

Humans make use of all five of their senses in an effort to communicate with one another. But things are a little more complicated than simply using sight, smell, taste, hearing, and touch. The fact that humans are capable of abstract and symbolic thought means that other, more sophisticated avenues of communication are possible.

Spoken and Written Language

Let us begin with perhaps the most obvious avenue of communication—the use of language. As was pointed out earlier, man's ability to speak goes back at least 200,000 years, and his use of the written word dates back some 5,000 years.

Scientists believe that the FOXP2 gene is the prime candidate responsible for human speech and language proficiency. The "Forkhead box protein P2" (FOXP2 gene) is located on human chromosome number seven. There is evidence that this genetic mutation goes back at least 200,000 years, and that not only do Homo sapiens possess the gene, but Neanderthals did as well. Besides providing the neurological sophistication for speech, FOXP2 is also associated with fine oro-facial movements. Modern humans' supralaryngeal airway allows for a better quality of non-nasalized sound, and is entirely different from any other terrestrial mammal. However, because of the unique location and positioning of the human larynx, we are more likely than other mammals to choke when eating.

Man's tremendous capacity for speech and language does enable him to communicate efficiently and effectively. According to a 2007 study done by University of Arizona professor Matthias R. Mehl, both men and women average about 16,000 words each day. That is about 1,000 words for each hour awake.

Most adults have a vocabulary of over 10,000 words, and many adults can speak several languages. The term "polyglot" refers to an individual who can speak many languages—in rare cases, a dozen or more.

Baby Talk

Right from the very beginning of their lives, humans try to communicate verbally. Immediately following birth, a human baby will cry, and a parent will respond by trying to comfort him. In fact, the adult response to a baby's "noise-making" is laying the foundation for the infant's language development.

At two months old a baby will respond to adult "sweet talk" by "cooing" in return. The infant is attempting to communicate. Most toddlers will utter their first words around one, and by around two years of age they start to understand and formulate complex sentences.

Dr. Anne Fernand, a developmental psychologist at Stanford University, has done considerable research on children's language development. "You need to start talking to them from day one," Fernand said at the American Association for the Advancement of Science annual meeting in Chicago. She added, "You are building a mind; a mind that can conceptualize, that can think about the past and the future." Indeed, studies show that infants whose parents do not make an effort to talk with them lag behind by as much as 6 months in their language skills by the time they are two years old. (*The Guardian*, Feb. 14, 2014)

> *"What makes us human, I think, is an ability to ask questions,*
> *a consequence of our sophisticated spoken language."*
> *- Jane Goodall (British primatologist and anthropologist)*

All this verbal communication enhances our ability to cooperate and collaborate. We can tell someone how to make a fire, how to build a shelter, or where to find food. We can tell someone how to do a math problem, how to do a science experiment, or how to build a bridge. We can tell people that we are sad, that we are angry, or that we love them. While not as efficient as the Borg's use of mental telepathy, words do allow us to share what we are thinking and feeling fairly effectively.

The Ability to Write

Besides oral communication, man has developed a system of writing. Evidence suggests that some of the earliest forms of writing from ancient Egypt, Mesopotamia, and India go back more than 5,000 years. There are indications that Chinese writing goes back about 3,500 years, and that Mesoamerican writing dates back about 2,500 years. The invention of writing added a whole new dimension to man's ability to communicate.

The written word allows us to do several very important things. Whether the writing is on tree bark, modern paper, a computer screen, or on any other number of electronic devices, writing enables us to communicate over great distances and to large numbers of people.

The written word enables us to communicate not only with our contemporaries, but also with people in the future. For example, the Torah and Christian bibles were written thousands of years ago, but are still read today, and probably will be read for some time to come. And so we can pass on moral, philosophical, and scientific information from one person to the next, from one country to the next, and from one generation to the next.

Not only does writing allow us to convey thoughts and information, but it also allows us to standardize that information so that the dissemination of that information does not distort the original content. Everybody understands that passing on information verbally—from person-to-person, or generation-to-generation—is problematic in that the information or story often changes over time.

Most people are familiar with the game "telephone," where a message is passed verbally from one individual to the next, through a series of individuals. By the time the message reaches the last in a long line of participants, the original message has changed quite a bit. On the other hand, passing a written note through a series of people would not change the content of the original message.

The written word helps us communicate with many people, over great distances, over great periods of time, and do so more accurately than the spoken word. The invention of writing helps man communicate, cooperate, and collaborate more efficiently and more effectively, and therefore enhances his capacity to socialize in very complicated and elaborate ways.

So fundamental is language to our humanity that we manipulate it to form poems and set it to music. So important is language that we put it on paper, spray paint it on walls, chisel it on stone, and give it wings with modern wireless communication devices.

"Communication leads to community, that is, to understanding, intimacy and mutual valuing." – Rollo May (American existential psychologist)

Without adequate communication group solidarity, unanimity, and harmony will inevitably suffer. Our linguistic ability is one of several attributes that help keep us together. Besides language, humans communicate with each other in a host of other ways.

Body Language

Like many other animals, humans can communicate non-verbally through body language. Body language includes the use of gestures, posture, facial expressions, touch, and even proximity.

Facial expressions can involve the use of the eyes, eyebrows, lips, and cheeks. These facial movements can be used to convey moods such as happiness, sadness, depression, or anger. The importance of our eyes in the communication process was investigated by the Max Planck Institute in 2002. When compared to gorillas and chimpanzees, human infants followed eye movements better and more often than our primate cousins. No other primate has a sclera (large white part of the eye). Scientists hypothesize that having a sclera makes it easier to track eye movements, thereby enhancing the communication process.

Body posture, or positioning, is also meaningful and can convey many things. For example, sitting with your arms folded and your legs crossed while someone is talking to you could indicate your lack of interest, or even a degree of closed-mindedness. On the other hand, sitting forward in your chair and nodding your head would probably suggest that you are focused and paying close attention to the speaker. Standing akimbo (hands on hips with elbows flung outward), facing the speaker, is also usually interpreted as being interested in what the speaker has to say.

Gestures with your arms or hands can also be used to communicate. Clenched hands usually indicate stress or anger, while relaxed hands suggest confidence and self-assurance. Some gestures can vary from culture to culture. Giving the "thumbs up" is a positive gesture in most Western countries, but is viewed very negatively in Iran, Bangladesh, and Thailand. (It is almost the equivalent of using the middle finger in Western culture.) Similarly, handshakes between men and women are acceptable in Western societies, but are frowned upon in many Muslim countries. Even proximity, or distance between individuals, can communicate intimacy or dissatisfaction.

Humans are very good at reading body language (kinesics), and like written and verbal language, body language aids and enhances interpersonal communication.

Tears and Empathy

One uniquely human behavior that perhaps can be categorized as a type of body language is the production of human tears and silent crying. Tears can play an important role in communication. The extraordinary thing about tears is that they don't just telegraph our state of mind to others, they can also evoke strong emotions in the people who witness them. One theory is that crying may have evolved as a signal that was valuable because it could only be picked up by those physically closest to us—who could actually see our tears. Tears communicate fear and distress to fellow group members who are close by (who would be more likely to help), and at the same time, more distant predators wouldn't know that you are vulnerable. Our ancient ancestors' ability to evoke empathy in fellow group members promoted a sense of security and support that helped build harmony and fraternity within the community.

The da Vinci Within

We are all familiar with the adage that "a picture is worth a thousand words." Long before man was writing things down, he was painting figures on cave walls, carving figurines, and making jewelry. Necklace fabrication has been dated as far back as 80,000 years, cave paintings and other petroglyphs go back about 40,000 years, and figurines date back approximately 35,000 years. What is the evolutionary purpose of art? Just like language, its evolutionary function is to facilitate communication. This is not to say that art couldn't have some secondary purposes and benefits—like being therapeutic. But enhancing communication is its primary function.

"I found I could say things with color and shapes that I couldn't say any other way— things I had no words for." - Georgia O'Keeffe (American artist)

Paintings on cave walls, for example, can convey all sorts of important information. They can tell the viewer about what animals are in the area, how to hunt those animals, and which animals to be wary of. They can convey information about the tribe's belief system. For instance, some ancient art suggests that certain animals had a spirit to be reckoned with, or that the sun was an object to be worshipped. Cave drawings tell fellow clan members about their culture, their history, and who they are as a people. One can perhaps see similarities with today's graffiti artists whose works tells others who they are and where their defined territory is. This is important information that helps define group association and helps keep groups bonded together.

Evidence of shell necklaces goes back even further than petroglyphs. One such necklace was found in Grotte des Pigeon cave in northeastern Morocco and dates back some 50,000 years. Such types of adornments, like jewelry or body painting, serve to strengthen and confirm group membership. A necklace could identify the wearer as being a member of a particular clan or tribe and signify the protection and acceptance that goes along with group membership. It might also be a signal for the possibility of some sort of inter-group trade, or maybe the opportunity for mate selection and exchange.

Of course, group identification can be problematic when encountering members of a competing group. In today's society, wearing a hat or jacket with a sports team logo on it might gain you acceptance with like-minded fans, but it also could present some problems if you are in the midst of fans of an opposing team. Along the same lines, wearing gang colors or displaying a gang logo on your jacket helps solidify gang membership and enhance group cohesiveness, but there might be negative consequences in terms of inter-gang conflict. In general, however, these kinds of artistic symbolizations do help keep groups well-defined and bonded together.

We can go even further back with artwork by considering the fabrication of stone tools. The manufacture of stone tools goes back over one million years. There are indications that the creation and design of some of these tools went beyond their utilitarian value and were artistically creative in nature. The aesthetic nature and design of such stone tools communicated not only group membership, but also might have had the additional purpose of being a demonstration of the knowledge and skill of the creator. This could have had implications as far as mate selection goes, and would have clearly demonstrated tribal attributes and prowess to strangers.

A modern example of such behavior might be the United States landing a man on the moon in 1969. Americans felt proud and connected to their nation (group). At the time, we probably did not fully understand all the technological benefits that would spin off as a result of this feat. But we were clearly sending a message to the rest of the world about who we were as Americans, and what we were capable of.

> *"Art is not a handicraft, it is the transmission of feeling the artist has experienced." - Leo Tolstoy (Russian writer)*

Besides being factually informative, art can also have emotional content. Art can communicate a whole range of human emotions including happiness, sadness, anger, fear, anxiety, love, and hate. Paintings, for example, can have joyous scenes and subjects with happy faces. Or the painting can be dark and forbidding. There is a famous expressionist painting entitled "The Scream," by Edvard Munch (1893). In this painting the facial expressions and body language of the subjects, along with the scenery and setting, readily communicate pain, fear, and anxiety.

Another famous painting, done by Leonid Afremov (2000), and entitled "Dance Under the Rain," has as its subject an embracing couple obviously in love. There are bright lights and colorful trees. Vivid colors and open landscapes often evoke a feeling of relaxation and happiness, while images that are dark and obscure typically elicit anxiety and fear.

The mere choice of color can have an impact. Red, orange, and yellow are considered "warm" colors and often evoke feelings of optimism, happiness, and energy. Red is considered the warmest and most dynamic of the colors. Red can increase your heart rate and make you feel excited. It is a color that is often associated with passion, lust, sex, energy, anger, blood, and war. Yellow is often associated with sunshine and can elicit feelings of optimism and hopefulness. Green, blue, and purple are said to be the "cooler" colors. Green is easiest on the eyes and can help a person feel more relaxed. Blue evokes feelings of calmness and spirituality as well as security and trust.

It should be noted that some of these color associations could be somewhat different depending upon cultural differences. For example, black is perceived in most Western cultures to be associated with negative things, such as evilness and death, while white symbolizes purity. But in some Eastern cultures, white can be associated with death. While Westerners may choose to wear black clothing at a funeral, some ancient Chinese cultures preferred white clothes and hats at funerals.

Another artistic variable has to do with pattern recognition. Symmetry is often found in works of art, and the human brain unconsciously searches for symmetry for a number of reasons. Potential predators were bilaterally symmetrical as were potential prey. Bilateral symmetry also exists in humans. A healthy human is typically symmetrical. This attraction to symmetry was therefore advantageous, as it helped humans recognize danger, food, and quality mates.

Suffice it to say that art can be used to convey both factual and emotional information. As such, it facilitates communication and enhances human group cohesiveness. And even though everybody possesses different levels of artistic talent, artful expression is indeed ubiquitous. We all practice and understand it to a greater or lesser degree. And while there are some examples of non-human artwork (chimpanzee, gorilla, elephant, and bird), when it comes to artful expression, humans are in a category by themselves.

The Music Man

Is man's interest in music also one of our species' adaptive behaviors in response to natural selection? If so, what part does music play in man's effort to survive as a group—as a species?

Man's captivation with music is virtually universal and historic. Evidence of man's predisposition to music goes back about 40,000 years with the discovery of a prehistoric flute in Geibenklosterle cave in southern Germany. As far as we can tell, music has been part of every culture, in every corner of the globe, since man's beginning.

Human beings listen to or play all sorts of music. There is jazz, blues, rock, folk, ethnic, country, religious, rap, hip hop, classical, and heavy metal—to name a few. We listen to live performances, vinyl records, tapes, CDs, and MP3 players. We listen to the radio, podcasts, music on the Internet, and music on TV. Most people around the world have access to several of these sources, even in so-called "third world countries."

According to a recent study (Edison Research, 2014), Americans listen to slightly over 4 hours of music each day. That equates to about 2 million songs and roughly 120,000 hours over the average lifetime. One would suspect that in most modern, industrialized countries music involvement is not that different from that in the USA.

One in five people play some sort of a musical instrument, and according to Lawrence Rosenblum, a professor of psychology at the University of California, 85 to 90 percent of adults can carry a tune (pitch, rhythm, and melody) correctly. And our proclivity for music starts at an early age. According to psychologist Marcel Zentner of the University of York (England), recent studies with infants as young as five months old show babies responding to the rhythm and tempo of music by moving their arms, legs, and torsos.

Another example of human infants responding to pitch and rhythm has to do with "baby talk." Baby talk, or "Infant Directed Speech" (IDS), typically involves talking to an infant using a higher-pitched voice and a simplified vocabulary. For example, using "boo-boo" instead of wound, "choo-choo" instead of train, and "wuv" instead of love. IDS also involves using a wide range in pitch variations and a slower, more deliberate vocal rhythm. In 1990, researchers Robin Cooper (Virginia Polytechnic

Institute) and Richard Aslin (Virginia State University, Blacksburg) found that infants as young as two days old responded better to IDS than to normal adult speech. In a similar study (2011), researchers Adena Schachner (Harvard University) and Erin Hannon (University of Nevada) found that five-month-old babies responded more positively to strangers using IDS than they did to hearing their own parents speaking in a normal adult manner. What these two studies, and others like them, show is that humans appear to be hard-wired to distinguish variations in pitch and rhythm in both talk and song. Music is "in our blood."

> *"Music is the universal language of mankind."*
> — *Henry Wadsworth Longfellow (American poet)*

What evolutionary purpose does man's predilection for music serve? Let us remind ourselves what this chapter is about. We are focusing on those innate human characteristics and behaviors that help facilitate human group bonding and functioning.

To begin with, music helps us communicate, and we know that good communication is essential for group functioning. Music can be informative. Because melody and rhythm come so naturally to human beings, music can be used to help youngsters memorize the alphabet, learn the multiplication tables, the months of the year, and historical facts and dates. Songs can be used to help teach a foreign language, to teach tolerance, or to teach people about health and environmental issues. Folk music and ballads can teach us about morality and social justice. It also so happens that the two great literary pieces, *The Iliad* and *The Odyssey*, started out as chants.

Business owners and marketing people understand very well that making up a catchy musical jingle can be very helpful to get people to remember their product. One of the most successful commercial tunes of all time was the 1971 Coca Cola song, "I'd Like to Teach The World to Sing." It went on to be a top pop song at the time. And many people—especially baby boomers—can remember the catchy little melody and lyrics for Wrigley chewing gum. "Double your pleasure, double your fun, with Wrigley Doublemint, Doublemint gum."

Music can also be a very effective tool for the communication of emotions. People often say that they can "feel" the music. Humans can perceive emotional content even if the lyrics are in a foreign language, or even if there are no lyrics at all. Indeed, melody and rhythm are a language in their own right. Songs can make us happy, or melancholy, or relieve anxiety. Music can evoke feelings of spirituality and the divine. There are songs in the Hebrew Tanakh, the Christian Bible, the Quran, and the Bhagavad-Gita. People sing in temples, churches, and mosques. Christian monks used to chant for hours on end, and slaves sang songs to help relieve the monotony and drudgery of hard labor.

Music can stir us up. Music in parades can elicit strong feelings of national pride and patriotism. And let us not forget that the beat of the drum and the sound of the bugle lead many a soldier into battle.

All of us have experienced situations in which hearing a special song has brought back memories. Music can throw us back to a certain time and place and refresh our memories about people and situations. In fact, researcher Petr Janata, at UC Davis Center for Mind and Brain, discovered a hub in the brain that links music, memory, and emotion (Liese Greenfelder, "Study Finds Brain Hub That Links Music, Memory and emotion," *ucdavis.edu.news,* Feb. 23, 2009).

Music can bond us with friends and groups. Singing together or listening to the same music can give us a sense of togetherness, solidarity, and a feeling of harmony with those around us.

Music isn't some curious human endeavor whose incidental purpose is to amuse and relieve boredom. It isn't some accidental, innate human proclivity that nature has serendipitously bestowed upon us. Music is in fact one of many adaptive behaviors that serves to enhance man's group cohesivity.

It is interesting to point out that humans do not rank very high when compared to other terrestrial animals when it comes to a sense of smell. A dog's sense of smell is over one million times more sensitive than man's. Black bears can smell food over 10 miles away, and a polar bear can smell a seal through 3 feet of ice. Many animals communicate through odors and a good sense of smell. Perhaps man's sense of music—melody and rhythm—helps make up for a rather mediocre sense of smell. Music is certainly an additional channel of communication.

Before we leave the topic of music altogether, let us also briefly look at the role that dance plays in man's evolution.

Dance Fever

Let us start with the observation that just about everybody—no matter how old they are, no matter what corner of the Earth they are from—dances from time to time. And although some individuals are more adept at dance than others, we all do it, and we do it over and over again throughout our lives. Some of us are specially trained to dance, but no special training is required. Where there is music and rhythm, feet will be tapping and bodies will be swaying. This being the case, we are drawn to the conclusion that man's penchant for dance is another one of those evolutionarily derived behaviors.

"Dance is the hidden language of the soul"
— Martha Graham (American modern dancer and choreographer)

Hard evidence for Paleolithic dancing is difficult to ascertain. Signs of prehistoric dancing do not endure for a million years like stone tools do. There are 9,000-year-old paintings of dancing figures in India's Bhimbetka rock shelters. And there are 6,000-year-old Egyptian tomb paintings of dancers. But nothing has been found that goes farther back than that. We can, however, gain some additional insight into prehistoric dance by relying on research and ethnographic studies that have been done on some primitive tribal societies that have persisted into modern times.

What theses studies of primitive peoples show us is that dance can be used to communicate many important things. It can be informative, spiritual, and can convey emotion. There are dances for weddings, funerals, and festivals. There are welcome dances, love dances, and dances to signify various rites of passage. There are dances to signify the beginning of war, and dances to communicate with the spirit world. Perhaps a case can be made that dance, as a type of body language and as a form of communication, predates speech.

Just as important as its communicative value, dance is a bonding experience. In a study (Jason Goldman, "Why Dancing Leads to

Bonding," *Scientific American*, May 1, 2016) done by University of Oxford psychologist Bronwyn Tarr, Brazilian high school students were asked to dance—some in sync, some out of sync with their partners. Participants said they felt closer to their partners when they danced in sync (same steps at the same time) than when they danced out of sync. They also reported feeling closer the more they exerted themselves while dancing.

Moving in concert with other human beings is innately gratifying. It involves the release of endorphins (pain and stress reducing neurotransmitting chemicals). Whether it is a couple dancing together for the first time, or a group of people moving synchronously to the rhythms—it is a shared experience, and one that offers a sense of belonging and solidarity with your fellow dancers. It is neurologically rewarding.

"Our biological rhythms are the symphony of the cosmos, music embedded deep within us to which we dance, even when we can't name the tune."
— *Deepak Chopra (Indian American author – New Age Movement)*

Both singing and dancing are examples of behaviors that enhance connectivity, and promote cooperation and social harmony. It is no accident that that we can comfortably use the musical terms "harmony" and "orchestrate" when referring to music's positive effect on group cohesiveness.

Religion and Spirituality

Although there will be a subsequent chapter devoted to the topic of religion, let us focus here on how human religiosity can be a force to help bind people together. Let us begin with the fact that some form of religion or spirituality is practiced in every corner of the globe, and has been since our species' very beginning. No matter what time period we look at, no matter what area of the planet we observe, the vast majority of humans are demonstrating some form of religiosity. It is ubiquitous, and should be considered innate, just like language, music, and art.

How does religion serve group cohesion? Religious practices involve rituals and ceremonies—lots of them. They involve people coming together to bear witness, to pray, to sing, to dance, or to share a sacred meal. People who identify with a particular religion adhere to a common set of beliefs. They believe in the same god, gods, or spirits. Generally speaking, they accept the same worldview, and make an effort to live by the rules, laws, and moral code prescribed by their chosen faith. Religion represents a kind of "ideological glue" that helps a group of people stay connected and bonded.

In ancient times various tribes worshipped their own particular god or gods. For example, Yahweh was one of the gods of the ancient Israelites, Zeus was one of the gods of ancient Greece, and Aten was one of the gods of ancient Egypt. A group's god helped define the clan, ethnic group, or nation. Having the same god helped to identify an individual as a member of a particular group. And although some people may have had some inner doubts about the existence of their deity, they would have been reluctant to express these reservations publicly fear of being rejected by their fellow group members. Indeed, our human need to belong and be accepted is quite strong.

All of this connectivity and togetherness in body, mind, and spirit serves to help an individual identify with the other group members. Everybody feels connected and a part of the whole. Religion helps to satisfy that human urge to group affiliate, and serves to bind the group together. Like language, music, art, and dance, religiosity is part of our human repertoire, and is still another example of a behavior that is evolutionarily purposeful.

Leadership and Followership

Let us consider one last pair of human behaviors that facilitates group integrity and group cohesion—and that is "leadership" and "followership." From Paleolithic times to modern times, every functioning group has had both leaders and followers.

It is quite natural for some sort of a leader to emerge whenever humans get together to form a group, and quite natural for the others to "follow the leader." Let us look at leadership first.

Leadership

Not only is it instinctual for a group to want to have a leader, but there is also a Paleolithic urge to want to have a strong leader to boot. Ancient hunter-gatherer units were most likely initially led by the founding male and female genetic pair. As a clan grew in number, successive leaders would be established based on valued traits. Perhaps it would be the biggest, boldest, fiercest individual. Maybe it would be the best hunter, or perhaps it would be the smartest. In any case, humans know a good leader when they see one.

Ethnographic studies of contemporary (or near contemporary) hunter-gatherer peoples do not reveal any evidence of "democratic" urges when choosing a leader. Hunter-gatherer groups do not select a leader by a "show of hands." And even though modern democracies do select their leaders by voting in countrywide elections, there is some evidence that even in modern times, people continue to look for a "strongman" as their chosen leader—both in politics and in business.

For example, in U.S. presidential elections, the taller of the two candidates has been elected over twice as often as the shorter candidate. And although only about 2% of Americans stand 6'2" or taller, Malcolm Gladwell, in his book *Blink,* found that 30% of Fortune 500 CEOs are at least that tall.

Further evidence of this prejudice was revealed in a survey conducted by Dr. Gregg Murray at Texas Tech University in Lubbock. Dr. Murray, whose work was published in the journal *Social Science Quarterly* (2011), asked 467 students at American universities to describe and draw their "ideal national leader" alongside a "typical citizen." Almost two-thirds (64%) of the participants drew leaders who were taller than their average citizen. The leaders were on average 12% taller. While most of those who took part in the study were American, students from Africa, Asia, Europe, and Latin America were also involved. According to Dr. Murray:

"It would be possible to argue that this effect might be cultural. However, we found the results held across different cultures, suggesting an evolutionary cause. Our ancestors lived in groups that were constantly engaged in conflicts that were resolved through physical violence. If you are in a group and the

enemy hordes are coming over the hill, what you want them to see is the big person out front so they know they face a tough battle."

Height also translates into earnings. Columbia management professor Sheena Iyengar has found that men earn about 2.5% more per inch of height. In addition to height, researchers at the University of California at San Diego and Duke University (Mark van Vugh, *Psychology Today,* 2013) reported that men with deeper (lower) voices earned $187,000 more annually than their counterparts. (Deeper vocalizations are in general associated with larger, stronger animals.)

In the business world, firms and corporations have managers, directors, and company presidents. The military is similar with its hierarchy of leadership positions. There are sergeants, lieutenants, colonels, and generals. In all of these scenarios, the person in charge is expected to possess and display a variety of leadership qualities, but there is still a hint of a Paleolithic urge to value physical prowess, or bravado, when it comes to selecting a leader. And let us not forget that many nations throughout the world are not true democracies anyway, and are instead led by military-style strongmen.

Looking Up to the Leader

Humans seem to not only desire a strong leader, but they also seem quite ready to bestow upon that leader an elevated status—a status beyond that of an ordinary man or women.

For example, during the Middle Ages certain family lines emerged and came to dominate a particular geographical area. Many times these individuals and their families acquired power through intimidation and warfare. Some of these bloodlines evolved into full-blown monarchies, headed by a so-called king and queen whose claim to the throne was based on the idea of divine right (God's will). As such, these "divinely chosen" people enjoyed a kind of quasi-divine status and were considered to be a manifestation of or an agent of God. If we go a little bit further back in time, some rulers, such as the Egyptian pharaohs, were even considered to be a god by their subjects.

People today are still fascinated by the idea of a "royal family."

Thousands stand in line to get a glimpse of Queen Elizabeth or some other members of the British royal family. We sometimes refer to such royalty as "blue bloods," a term that seems to imply some sort of a genetic basis to their aristocracy. One might wonder if our genetic makeup has anything to do with a capacity for leadership.

> *"Leadership is something I was born with."*
> *- Carmelo Anthony (American professional basketball player)*

According to Ron Riggio, professor of Leadership and Organizational Psychology at Claremont McKenna College in California, "Leadership is hardwired into our DNA, and it comes from dominance and social hierarchy in animals." Various studies do support the idea that leadership can be rooted in an individual's nature.

There is evidence that genetic differences are correlated with the likelihood that certain individuals will take on managerial responsibilities. A study done in 2006 and 2007 by Dr. Richard D. Arvey at the National University of Singapore indicated that identical twins had a higher coincidence (31%) of leadership traits than did fraternal twins.

Similar results were discovered in research done at University College, London. This study, published online in *Leadership Quarterly* (2013), was the first to identify a specific DNA sequence associated with the tendency for individuals to occupy a leadership position. Using a large twin sample, an international research team which included academics from Harvard, New York University, and the University of California estimated that genes passed down from their parents can explain about 25% of the observed variation in leadership behavior between individuals. Other similar studies have the correlation as high as 40%. And so it seems that our genes do play a role in behaviors associated with leadership.

Followership

It can be argued that down deep everybody understands that groups of people, especially large groups of people, need to coalesce around a leader—or nothing will get done. High-functioning groups not only have a good leader, but they also have cooperative followers. Both are equally important. Having established that certain individuals will step forward to assume a leadership role, let us now look at how willing others are to follow. Has human evolution seen to it that the human species contains an effective mix of leadership and followership qualities?

In a way, "followership" is the default setting in our brain, and it begins when we are very young. According to Mark Van Vugt, a Professor of Organizational Psychology at VU University in Amsterdam, human predilection for followership begins almost at birth.

> *"Within minutes of birth, babies start mimicking the facial expressions of their mothers and from the age of about three months they will follow the mothers' eye gaze. From about nine months on, children will look from the object that their mother is gazing at back to the mother to check that they are both looking at the same thing. Then from 14 months onwards they are able to direct the gaze of their mother to an object, for example by pointing, so that they are coordinating their activities. The mother-infant relationship is therefore the first form of leadership-followership that we humans encounter, and it is a survival strategy."*

In their book *Why Some People Lead, Why Others Follow, and Why It Matters,* Dr. Mark Van Vugt and Anjana Ahuja bring to light an additional point. Many animal groupings, such as a wolf pack or a lion pride, revolve around a dominant individual. In most non-human primates the leader is the dominant male in the troop—the alpha. In gorillas, the secondary individuals are attuned to the actions of the alpha, the silverback. When he moves, they follow. The focus on the alpha serves to foster group cohesion, which is critical to the survival of all primate groupings.

The same phenomenon exists in humans. Boy scouts gather around their troop leader. Workers follow the dictates of their foreman. Soldiers unite around their sergeant, lieutenant, or captain. Britons rally around their queen. Humans accept hierarchy quite readily and naturally, which is essential for group cooperation and functioning. In general, people would like to be proud of their group leader. They want their leader to be a good example (specimen) of their group—someone they are willing to follow.

*"The world is moved not only by the mighty shoves of heroes, but
also by the aggregate of the tiny pushes of each honest worker."*
- Helen Keller (Deaf-blind American author and lecturer)

Additionally, followers learn from leaders, and at some point in time a follower could be called upon to assume a position of leadership. After working with and observing your foreman for several years, you may someday become the foreman. This is the way it has been through the millennia, a successful combination of human leadership and followership. Humans "get" how a group is supposed to function.

Followership Gone Too Far

So natural is human followership that it can sometimes go too far. By too far we mean that people sometimes follow a leader—often a charismatic one—blindly, and very unfortunate events occur. Consider the following rather famous examples of abusive leadership coupled with blind obedience:

David Koresh was the leader of the Branch Davidians who inhabited a compound in Waco, Texas. Rumors and later reports from ex-group members suggested Koresh had married several members, some in their mid-teens, and physically and sexually abused them. A 1993 FBI raid on their compound left 76 dead, which more or less resulted in the eradication of the group.

Marshall Applewhite was the leader of the Heaven's Gate sect, based in California. In March of 1997, thirty-eight members took their own lives with the promise that their suicide would allow them to shed their bodily "containers." They were told that after their demise they would be able to hitch a ride on a spacecraft hidden behind the comet Hale-Bopp to reach a higher existence.

Warren Jeffs was the leader of the Fundamentalist Church of Jesus Christ of Latter Day Saints (which broke off from the Mormon church in the 1930s over the issue of polygamy). The allegedly polygamous group's Texas compound was raided in 2008. Authorities took into legal custody more than 400 children and 133 women—all deemed to have been harmed or in imminent danger of harm.

Let us consider one more example of extreme leadership and followership, one that we might consider a personality cult on a national scale. The Kim family has ruled North Korea since 1948. First there was Kim Il Sung, followed by Kim Jong Il, and presently North Korea (DPRK) is being ruled by Kim Jong Un. This country is marked by the intensity of the people's feelings and devotion to their leaders. Said devotion requires total loyalty and subjugation to the Kim family, resulting in a one-man dictatorship through three successive generations—so far.

The 1972 constitution of the DPRK incorporates the ideas of Kim Il Sung as the only guiding principle of the state and his activities as the only cultural heritage of the people. Kim Il Sung has been described as a god, and Kim Jong Il as the son of a god, or "The Great Sun of Life." There is even widespread belief that Kim Il Sung created the world and that Kim Jong Il controlled the weather. Kim Il Sung is credited with almost single-handedly defeating the Japanese in World War II. The bodies of Kim Il Sung and Kim Jong Il are on display in North Korea, each in a glass sarcophagus. It should be noted that similar displays have been done for Vladimir Lenin (Russia) and Mao Zedong (China). This kind of veneration reminds us of the efforts of the ancient Egyptians to preserve the bodies of their "god-king" thousands of years ago.

We can see from these examples how an individual's proclivity for leadership can sometimes take on cult-like or even god-like characteristics. And humans seem quite willing to imbue their leaders with superhuman qualities. Man's penchant to be a loyal and cooperative follower can make him susceptible to these larger-than-life personalities.

Let us end this train of thought by reminding ourselves that history is filled with dictators and dictatorships, and that humans have a Paleolithic desire for a strong leader. Often, in times of trouble, people look for a strongman as an anecdote to political and economic chaos. Indeed, for the most part, dictators enjoy a significant degree of support from the citizenry of their respective countries, especially initially, allowing each of them to come to power in the first place. The following list represents some rather famous examples of relatively recent dictators: Adolf Hitler, Josef Stalin, Mao Zedong, Benito Mussolini, Pol Pot, Kim Jong Il, Saddam Hussein, Robert Mugabe, Fidel Castro, Chiang Kai-shek, Idi Amin, and Francisco Franco—to name just a few. Each of these men had their loyal followers.

The combination of a charismatic individual with an over-sized ego and man's age-old instinct to seek and follow a strong group leader in turbulent times can lead to trouble. One can perhaps wonder if World War I and World War II—with a combined death total of about 100 million—could have been avoided if humans weren't so inclined to "follow the leader." This is a case in point regarding true human nature. *The Paleo Perspective* is a cautionary note, reminding us of our inherited proclivities, some of which have the potential for great harm.

Peer Pressure and Group Conformity

Our discussion of human followership wouldn't be complete without a brief mention of the phenomenon of group conformity or peer pressure. While a certain amount of "groupthink" can facilitate group cohesion and functioning, "conformity" comes with a warning label.

In 1932, psychologist Arthur Jenness performed one of the earliest experiments on the phenomenon of "conformity." Jenness asked participants to estimate the number of beans in a bottle. They first estimated the number of beans individually (separately), and then later did so in a group setting. Finally, they were once again asked, individually, for the bean count. Jenness found that estimates shifted from their original guess to one that was closer to what the other members of the group had guessed in the group setting.

There is also a well-know study conducted in 1951 on peer pressure by social psychologist Solomon Asch at Swarthmore College. The Asch conformity experiments were a series of studies published in the 1950s that demonstrated the power of conformity in group settings. These experiments consisted of a group of "vision tests," where "pseudo participants" (those who were working for and in collusion with the experimenter) first gave the actual participants incorrect answers.

The results were very interesting. It seems that 75% of the participants conformed to the incorrect answer at least once, and 5% conformed every time. When surrounded by individuals (pseudo participants) all voicing an incorrect answer, participants provided incorrect responses 32% of the time. Overall, there was a 37% conformity rate by subjects. In a control group, with no pressure to conform to an erroneous answer, only about 3% of the subjects gave an incorrect answer.

The experimenters also looked at the effect that the number of people present in the group had on conformity. When just one confederate (working for the experimenter) was present, there was virtually no impact on participants' answers. The presence of two confederates had only a tiny effect. However, the level of conformity seen with three or more confederates was far more significant.

We can see from these various studies that humans have a desire to conform, a desire be in unison with the group at large. This kind of followership, in conjunction with strong leadership, can be an effective combination to help facilitate group cohesion and group functioning.

We might also mention that in addition to the inclination to conform to the group's wishes, there is also a human desire to be recognized by the group for one's contributions to the group's effort. We often see this desire to contribute to the group's success on display in team sports or on the job.

Some Final Thoughts

As stated throughout, man is a highly socially complex animal. Evolutionary pressures have ensured that we humans have ingrained in our nature many different behaviors and traits that promote group connectivity and functionality.

Perhaps the most obvious one is the strong bonding forces that are in play when people are genetically related through their family lines. However, beyond genetic relatedness, our ability to communicate effectively in so many different ways is essential for group functionality.

Mentioned earlier was the fact that humans average about 16,000 spoken words each day. We read books, magazines, and newspapers. Nowadays we send e-mails, text messages, pictures, and videos. Many people (especially the young) are constantly on their cell phones, even while walking or driving. Whether it is the spoken word, written word, physical gestures, art, music, dance, or spirituality, we humans manage to express our feelings and make our thoughts known to each other. In fact, we spend the vast majority of our day communicating and sharing consciousness. While all of these avenues of communication are not as simple and direct as the aforementioned telepathic communication of the fictitious Borg, our level of communication and shared consciousness is second to none—at least on planet Earth.

Lastly, we should note that humans love to congregate and socialize. We love to get together with family and friends. Humans go to parties, festivals, concerts, and religious gatherings. And while there are specific attractions at these gatherings, such as music, dancing, or spiritual expression, the simple act of congregating and communicating is innately pleasurable. Recent research (Susan Weinschenk, "Why We're All Addicted to Texts, Twitter and Google," *Psychology Today*, September 11, 2012) regarding the almost addictive power of cell phones—be it talking, texting, or "googling"—alludes to the role of dopamine in a neurological "seeking and pleasure loop" operating in our brain.

Nature motivates and prods beneficial behaviors by making them pleasurable (e.g. sexual pleasure encourages procreation). Nature has encouraged human congregation and socialization by making it pleasurable. Human sociality is no serendipitous accident. Man has survived by being social. As pointed out earlier, solitary confinement is one of the most severe forms of punishment that a human being can experience.

"Nothing truly valuable can be achieved except by the unselfish cooperation of many individuals." - Albert Einstein (German-born theoretical physicist)

CHAPTER 12

THE FORCES THAT DIVIDE

"Small communities grow great through harmony, great ones fall to pieces through discord." – Sallust (Roman historian, 1ˢᵗ century B.C.)

We can see from the last chapter that there are many things that help keep human groups together and functioning well. With so many factors at work enhancing human harmony, one might wonder why there seems to be so much discord and conflict in the world.

There are, however, two significant factors that can often work against human harmony. One factor, which works against *intra-group* (within a group) solidarity, is group size. The other factor, which works against *inter-group* (between groups) cooperation, is social heterogeneity.

Group Size

Let us investigate group size first. Is there an optimum size for human groupings? While it is difficult to ascertain the exact group size for some of our ancestors such as Australopithecus, Homo erectus, or Homo heidelbergensis, most anthropologists figure these hominins lived in clan-type groupings that ranged from about 20 to 60 individuals. Estimates indicate that the number of individuals in ancient Homo sapiens groupings could have been similar, but may have been higher, with as many as 100 or more.

Psychiatrist Robin Dunbar, of Oxford University, believes that man's larger group size and more socially complex life style was made possible because of Homo sapiens' larger brain, specifically the neocortex. The theory is that we required greater neural processing power to enable us to keep track of who is doing what to whom, and why (social intelligence).

Back in 1993, Dunbar, along with anthropologist Leslie Aiello, crunched a bunch of numbers from the hominid fossil record and from observations of present-day apes. They found that the larger a species' group size, the larger the size of the brain, and in particular the larger the relative size of the neocortex (center for social intelligence) to the brain's overall size. Based on their findings they concluded that the correlations between group size, brain size, and neocortex size varied very little throughout our lineage, from australopithecines (3 million years ago) to modern-day humans.

From the study's data, Dunbar calculated that as far as modern-day humans are concerned, the number of people with whom the average person can comfortably maintain a close and stable personal relationship with is about 150 ("Dunbar's Number"). The implication of this study (and others like it) is that the sheer size of a group can put a strain on our human capacity to relate to and cooperate with fellow group members.

Shame and embarrassment, for example, can be very effective at mitigating errant behavior. Sympathy and empathy encourage cooperation. But these emotions work best in small, intimate settings. They lose their effectiveness when the grouping is large and impersonal. At some point a group can become too large to maintain the necessary level of intimacy required for solidarity and cohesiveness.

Most of us are familiar with how difficult it can be for large groups to function. Larger groupings require more structure, more rules, and more patience. The fact is that we evolved in group settings that probably consisted of 100 or fewer individuals. Humans need a certain level of intimacy for *intra-group* cooperation.

As our towns, cities, and states have become larger and more densely populated, man struggles to relate to and cooperate with his fellow human beings. As these groupings become larger and less intimate, we lose the sense that we are even part of a group and begin to consider others as competitors instead of cooperators. Disorder, dissension hostility, and even war often result. It is perhaps somewhat ironic that as human societies have gotten larger, our sense of community has diminished. People can even feel intensely alone on a crowded urban sidewalk. We are simply out of our social comfort zone.

> *"The minute your group gets so big you don't know anybody*
> *in it, and they don't know you, there's hell to pay."*
> *— Dorothy Canfield Fisher (American author and social activist)*

Group Heterogeneity

Having acknowledged that group size can be problematic to *intra-group* functioning and solidarity, let us consider the sources of *inter-group* conflict—conflict between differing groups within society as a whole.

Groups are delineated by a particular trait, feature, or characteristic that its members have in common. The features that help members identify with the group also serve to identify group members to outsiders. Herein lies the potential problem.

A clan or a tribe would be an illustration of a particular grouping, one that uses genetic relatedness (kinship) as one of its key defining features—perhaps its most salient feature. Modern-day examples would include Native American tribes, Jews, Romani (The term "gypsy" is somewhat pejorative.), and Tinkers (sometimes called Travelers). Within each of these groups there is a presumption of a certain degree of genetic relatedness amongst its members. And in order to help maintain that genetic relatedness, these groups sometimes discourage marriages with outsiders.

People might also be identified with a particular religion (e.g. Christian, Muslim, Sikh, Hindu), nationality (e.g. Italian, German, Swedish), or race (e.g. Caucasian, Negroid, Asian). One might be identified as a Northerner or a Southerner. One might be identified by one's language such as Russian, Cantonese, or Spanish. Although each of these identifying traits helps to keep that particular group unified, group identification can be problematic in society at large because most, if not all, of these groups are imbedded within a larger heterogeneous social context.

For example, a person could be a black man living in the United States among whites, Asians and Hispanics. An individual can be a member of the Navajo tribe, but that person also exists in the context of the multicultural, multiracial United States. Similarly, one can be a Jew or a Muslim living in predominately Christian France.

Quite often minority groups find themselves segregated or ghettoized. For instance, Romani tend to live in pockets throughout Europe, and American blacks tend to be localized in inner city areas. This separation may be due in part to the fact that sometimes minorities freely choose to live among their ethnocentric cohorts and enjoy the culture they have in common. But it could also be a result of a degree of prejudice and shunning by non-group members in the surrounding society.

A person who speaks Spanish probably finds it somewhat natural to identify with other Spanish speakers. But non-Spanish speakers might view Spanish speakers as outsiders and subject them to palpable scrutiny and suspicion. Similarly, Jews have historically been subject to intolerance and discrimination by the surrounding non-Jewish populations. This discrimination has resulted in several pogroms and even genocide (the Holocaust). Blacks in America were subjected to slavery and denied their civil rights by whites for centuries. Native Americans—who were considered savages and sub-human by white European explorers and settlers—were massacred, forced off their land, and placed on reservations. Regrettably, there are many more examples of bias and bigotry—race against race, religion against religion, and culture against culture. Too often one group finds itself in conflict with other groups, or in conflict with the surrounding majority.

> *"Prejudice is a burden that confuses the past, threatens the future and renders the present inaccessible."*
> *- Maya Angelou (American poet and civil rights activist)*

A History of Conflict

Inter-group conflict is not a recent phenomenon. We can see evidence of *inter-group* conflict as far back as 10,000 years ago. An article in *The New York Times* (James Gorman, "Prehistoric Massacre Hints at War Among Hunter-Gatherers," January 20, 2016) describes an archeological find on the shores of Lake Turkana in Kenya. The find involves the discovery of 12 relatively complete skeletons. Scientists believe that 10 of the 12 skeletons show unmistakable signs of a violent death. Partial remains of at least 15 other individuals were found at the site, and they too were thought to have died in the same attack. The evidence suggests that one group of hunter-gatherers attacked and slaughtered another, leaving the dead with crushed skulls, embedded arrow or spear points, and other devastating wounds. Human *inter-group* conflict does not appear to be new.

This "us vs. them" attitude that is encoded in our DNA could have had certain advantages. The group whose members had a greater sense of fealty toward their own kind was the one that would cooperate better, collaborate better, and would therefore have a better chance of surviving. Favoring your particular grouping over other groupings—being "prejudiced" if you will—does have some evolutionary advantages, but it can be problematic for *inter-group* cooperation.

As stated earlier, man evolved in small, genetically-related bands. And so not only can group size be problematic, but ethnic and cultural differences in larger, more complex societies can be problematic as well. We are most comfortable with people who share our genes, culture, religion, language, etc. This phenomenon is considered "ethnocentric" behavior, and it is how we evolved over the last several million years. This is part of human nature and has both advantages and disadvantages.

Modern societies confront man with large, heterogeneous, and complex groupings. These modern groupings strain man's sense of intimacy and community, and make it harder for him to sympathize, empathize, and cooperate with such a diverse set of cohorts. The same Paleolithic urges that help keep a well-defined group solidified can also cause conflicts between dissimilar groups. Our differences often serve to generate hostility between the various ethnic, racial, religious, and cultural segments of society.

We are genetically designed to be tribal and to be suspicious of those who we view as being different from us. It is a conundrum—a problem that must be acknowledged and dealt with in order to reduce the social consequences of these Paleolithic instincts. Perhaps Dr. Martin Luther King said it best when he called for "a worldwide fellowship that lifts neighborly concern beyond one's tribe, race, class, and nation."

The Big Picture

With this problem in mind, let us look at the extent of hostilities throughout history and throughout the world. Has there been conflict or war somewhere on Earth at every given point in time?

Chris Hedges of *The New York Times* looked into that very question. In his article (July 6, 2003) "What Every Person Should Know About War," he states: "Of the past 3,400 years, humans have been entirely at peace for only 268 of them, or just 8% of recorded history." Based on that statistic it seems like the history of civilization is a history of conflict and war, only briefly punctuated by periods of peace.

One might wonder if man's propensity for conflict and war will someday lead to his final destruction. That is certainly a possibility. In pondering such questions one might wonder why evolution hasn't weeded out such wide-spread *intra-species* violence.

But the sad fact is that the death of thousands—even millions— of people represents but a fraction of the world's total population, and is not large enough to put the survival of the entire species in serious danger.

For example, approximately 60 million people died as a result of fighting during World War II, the largest of all the wars. However, the world's population at the time was about two and a half billion. As horrific the deaths of 60 million people were, they represented less than 3% of the world's total population at the time. Thinking cynically, one might even argue that murder and war are useful in that they act to slow down the planet's march toward ultimate overpopulation. At the very least we can acknowledge that humankind has endured despite its capacity for violence. However, as man's weapons of mass destruction get ever more deadly, there is no telling what the future might bring.

> *"My first wish is to see this plague of mankind,*
> *war, banished from the earth."*
> *- George Washington (First President of United States)*

Having said all that, let us try to put some of these things in perspective. Homo sapiens are a highly complex species. We simultaneously have genetic proclivities to be both selfish and altruistic, discriminating and accepting.

Let us think back to the analogy that was offered in our discussion on genetics. It was suggested that alphabet soup might be a useful analogy when thinking about one's genetic makeup. There are many, and sometimes contradictory, messages in that genetic "soup." At any point in time, and under varying circumstances, one genetic "message" might be closer to the allegorical soup's surface and come to dominate our actions, while other genetic urges might sink to the bottom and out of sight—unexpressed. Man has thousands of genetic "dos and don'ts" in his makeup. Human beings possess both the instinct to cooperate with their fellow humans, and also to be suspicious and hostile toward those who appear to be different.

Social cooperation is the mainstay and perhaps the cornerstone of man's success. Even with man's capacity for violence, he is still capable of feeling empathy for, and cooperative with, people whose religion, race, and culture are different from his own. People of varied backgrounds cooperate all the time on everyday things—things such as business transactions, traffic regulations, and waiting their turn on line. Many of us cooperate with even larger and more global concerns. We might participate in a recycling program to minimize pollution, donate to help feed hungry children around the world, or be concerned and active about possible global warming.

There are several notable large-scale examples of human attempts to cooperate across ethnic, racial, and cultural boundaries. One such example is the European Union (EU). The EU involves over two dozen countries, with over 500 million people, that have joined together in an effort to improve the lives of all those involved. EU policies aim to ensure the free movement of people, goods, services, and capital across the European continent. It enacts legislation for the purpose of enhancing justice, and maintains common policies on trade, agriculture, fisheries, and regional development. Unfortunately, very recently (2017), there has been some deterioration of EU solidarity—the exit of the United Kingdom.

Another example of global level cooperation would be the United Nations (UN). Some might argue that the UN is not terribly effective at maintaining peace throughout the world. However, nobody really knows what the world would be like without it. Whatever one's opinion is, the United Nations is a grand attempt at worldwide cooperation. Other examples of international cooperation are the World Health Organization

(WHO), the International Partnership for Energy Efficiency Cooperation (IPEEC), and the International Space Station (ISS).

However, reflecting on the central theme of this book, we must remember that it is much easier to function in smaller, intimate, homogeneous groupings—as our ancestors did. It is more natural and comfortable to relate to people who speak the same language, who share the same religious beliefs, and share the same culture as we do. Cooperating with people who are different from ourselves requires constant vigilance and effort. It is indeed a "heavy lift."

Invasions From Outer Space

There have been scores of films in which the Earth has been invaded by some alien species. *War of the Worlds* (1953), *Invasion of the Body Snatchers* (1978), *Independence Day* (1996), and *The World's End* (2013) are prime examples of this genre of film. In all of these films, human beings forgot about any racial, religious, or cultural differences they may have had with each other in order to defend themselves from these invading aliens. Any differences humans had with each other paled in comparison with the differences they had with beings from outer space.

In this fictitious scenario, all human beings would see themselves as belonging to the same group (human), and would therefore cooperate and collaborate with each other in their battle with the more foreign and alien group. Differences are relative. Such is human nature.

> *"As man advances in civilization, and small tribes are united into larger communities, the simplest reason would tell each individual that he ought to extend his social instincts and sympathies to all members of the same nation, though personally unknown to him. This point being once reached, there is only an artificial barrier to prevent his sympathies extending to the men of all nations and races."*
> *- Charles Darwin*

PART III
THE PALEO-LENS

Looking at the present through prehistoric glass

CHAPTER 13

PALEO SEX

"My body feels like it is asking to reproduce."
- Shakira (Columbian singer, songwriter, and dancer)

We have been discussing man's instinct to survive, both as an individual and in group format. While perhaps not as primordial as the instinct to survive, the instinct to reproduce is certainly essential to the success of any species. The ability to reproduce successfully is sometimes referred to as "reproductive fitness."

Living creatures use lots of different strategies for producing offspring, but most strategies fall neatly into the categories of either "sexual" or "asexual" reproduction. Asexual reproduction generates offspring that are genetically identical to the single parent. In sexual reproduction, two parents contribute genetic information to produce unique offspring. Sexual and asexual reproductions each have advantages and disadvantages. There are some organisms—aphids, slime mold, sea anemones, some starfish, and many plants—that do both. Humans reproduce sexually, so let us look at human sexuality in the context of animal sexuality in general, and in the context of our evolutionary past.

Most of us consider sexual behavior to be innate, instinctual behavior. While sexual behavior does vary greatly from species to species and from individual to individual within a species, sexual activity per se is ubiquitous.

Humans sometimes have a difficult time thinking about themselves as just another sexually active species in the animal kingdom. There are many jokes about human sexuality (e.g. men think with their "little head" instead of their "big head"). Perhaps humor allows us to more easily accept libidinous behavior and still consider ourselves to be highly rational beings.

It is, however, helpful to examine human sexuality in the context of sexuality in the animal kingdom. In particular, if we look at the sexual characteristics and behaviors of our two closest relatives—the larger, common chimpanzee (Pan troglodyte) and the smaller "bonobo" (Pan paniscus) chimpanzee—we can see many commonalities. (Chimpanzees share close to 99% of our DNA.)

For example, the gestation period for the common chimp is about 34 weeks, for bonobos it is about 33 weeks, and for humans it is about 38 weeks. On the average, all three species usually have a single offspring for each successful pregnancy. Both common chimps and bonobos reproduce successfully by around 14 years of age, and most human females are physically capable of having a baby by 14. The menstrual cycles for the three do vary a bit, with humans being the shortest at about 28 days, common chimps at about 35 days, and bonobos at about 45 days.

With regard to promiscuity, neither chimps, nor bonobos, nor humans are truly monogamous (i.e. sticking with one partner for life). There are of course exceptions to any rule, but all three species generally engage in sexual activities with multiple partners throughout their lifetimes. In addition, all three species partake in autoerotic behaviors from time to time, and all three engage in homosexual behavior. Of the three, bonobos tend to be the most promiscuous in terms of having multiple sex partners—regardless of age or gender. All three species do practice a degree of incest avoidance, especially between mothers and adult sons.

According to psychologist William Lemmon (of the Institute for Primate Studies in Norman, Oklahoma), "The female chimpanzee manifests most, if not all, of the indices of sexual arousal and orgasm that occur in women." A 1981 study (M.L. Allen and W.B. Lemmon, "Orgasms in Female Primates," *The American Journal of Primatology,* 1981) reported that sexual responses detected in female chimpanzees included transudate secretion, clitoral tumescence, vaginal thickening and expansion, hyperventilation, involuntary muscle tension, arm and leg spasms, clutching, facial expressions (e.g. low open grin, low closed grin, eversion of the lips, protrusion of the tongue), and a panting type of vocalization.

There are of course some differences in the sexual behaviors of humans and chimps. For instance, human males often prefer youthful partners, while male chimps prefer older females. And most human males do not shake a tree branch to signal sexual interest in a female. But as you can see, there are indeed significant similarities between chimps and humans.

Non-reproductive Sex

As previously pointed out, reproduction is necessary for the success of any species. But does that mean that sexual activity is engaged in exclusively for reproductive proposes? We know for certain that humans engage in all sorts of sexual activities (e.g. oral sex, masturbation) knowing full well that offspring will not result. We also know that in general, sex is quite pleasurable for humans.

"Sex is the most fun you can have without laughing."
- Woody Allen (American comedian and film maker)

We do have indications that sex is also pleasurable for other animals and that they also engage in non-productive sex. For example, bonobos have sex even when the female is not fertile, and Asian lions will copulate with multiple mates hundreds of times in one day—much more than is necessary for reproduction. There is also the fact that non-humans engage in homosexual behaviors.

Homosexuality

Non-human homosexuality is well documented in at least 500 species according to Bruce Begermihl (*Biological Exuberance: Animal Homosexuality and Diversity,* 1999). Homosexual behavior has been observed in marine birds, monkeys, gorillas, chimps, orangutans, penguins, cattle, and rams— to name just a few. It so happens that around 10% of rams refuse to mate with females, but do readily mate with other rams (Simon Levay, *Gay, Straight, and the Reason Why: The Science of Sexual Orientation,* 2017). Petter Boeckman, of the Norwegian Natural History Museum, goes even further saying nearly 1,500 species, ranging from primates to gut worms, have been observed engaging in same-sex behaviors. Homosexual oral sex has been observed in mammals as diverse as rats, fruit bats, horses, goats, dolphins, cheetahs, sheep, and cattle.

This is not to say that all of these animals practice homosexuality routinely and do not interact with the opposite sex at all. After all, heterosexual behavior is essential for the continuation of any species. But sexual pleasure is a reward for the repetition of both heterosexual and homosexual behavior. And there is also the theory that homosexual behavior can sometimes be beneficial in that it may reduce conflict within a species.

> *"Homosexuality is neither a sin, nor an anomaly or a disease. It is an evolutionary variation." - Abhijit Naskar (Indian neuroscientist)*

Trying to assess the prevalence of homosexuality within the human population is difficult. There is still reluctance on the part of individuals to self-indentify as being homosexual, especially in places where homosexual behavior is stigmatized. According to the "National Health Interview Survey" (2013), a little less than 3% of the US population self-identify as being gay. Similar results were reported by Seth Stephens-Davidowitz ("How Many American Men are Gay?" *The New York Times*, December 7, 2013). Using surveys, social networks, pornographic searches, and dating sites, he estimated that about 5% of American men were gay. Given the fact that homosexual behavior is most likely underreported, estimates that range from 2% to 10% are within reason.

Human homosexuality in the USA is often looked at as a "black or white" thing—you either are or you're not. Clinically speaking, that is not accurate. There are instances of human bisexuality. According to the Center for Disease Control (CDC), around 5.5% of women self-identify as being bisexual and around 2% of men do (2011 – 2013 survey).

For some people there is also a moral dimension to homosexual behavior. They consider it to be a voluntarily-chosen sinful lifestyle. But not many of us would rush to assign immorality and sin to acts of homosexuality by simpler, non-human forms of life. Are those aforementioned homosexual rams choosing a life of moral depravity? On the contrary, the fact that homosexuality also occurs in the non-human animal world adds weight to the argument that it is a natural phenomenon, and not a willful act of immorality.

Autoerotic Behavior

Another example of pleasure seeking behavior that does not result in reproduction is autoeroticism, or masturbation. Sexologist Havelock Ellis, in his 1927 book *Studies in the Psychology of Sex,* identified bulls, goats, sheep, camels, elephants, and porcupines as species known to practice autoeroticism. For instance, the female porcupine has been observed using a stick as a "vibrator," holding one end of a stick and walking around, straddling it, as it bumped against the ground and vibrated against her genitalia. Adelie penguins (and many other birds) rub their cloaca (genitalia) against a rock or some other object.

What about human self-gratification? According to an article that appeared in *The Journal of Sexual Medicine* (D. Herbenick, "Sexual Behavior in the United States," October, 2010), 5% of women (ages 25 to 29) engage in solo sessions more than 4 times a week, and 20.1% of men do. The gap between men and women narrows down for those who report masturbating multiple times a month, with 21.5% of women and 25.4% of men in that same age group. However, women are usually at least 10 to 15 percentage points behind men in all the other age groupings. Also, as both men and women get older they are less likely to report masturbating at all.

Human self-gratification can take on many other forms. Autoerotic asphyxiation, aqua-eroticism, sadomasochism, objectophilia, and erotic behaviors involving bondage and domination would be just a few examples. All of these forms of autoeroticism do not result in sexual or asexual reproduction, and one would be hard-pressed to conclude that they contribute to the ongoing success of our species.

The very least we can say is that animal sexual behavior, both human and non-human, can be pleasurable, and that not all sexual behavior has reproduction as its goal.

Human Sexual Attraction

Using our Paleolithic lens, let us look at sexual attraction that appears to be a result of our evolutionary journey. What do men and women instinctively find attractive in each other?

According to David Buss, an evolutionary psychologist at the University of Michigan, men appear to be more focused on the physical attributes of women, while women seem to be more focused on a man's potential to provide security for herself and any offspring. This makes evolutionary sense. Males are looking to impregnate females that possess youthful vigor and reproductive health to maximize their chances of passing on their genes to the next generation. Women, who are for the most part the primary caregivers of any offspring, are seeking males that can provide both physical and financial security for themselves and their offspring. In *paleo-times,* that might have meant a big man with good hunting skills. In more modern times, we might be talking about men who have higher social status and secure sources of income. But beyond this generalization, there are some specific attributes that both men and women appear to look for in a potential mate.

Facial Symmetry and Beauty

Many animals, including humans, look for physical symmetry in their mate. Symmetry turns out to be a fairly reliable indicator of good health. In one study (1998), David Waynforth of the University of Mexico measured the symmetry of men's faces in Belize and found that the men with more asymmetrical faces were more likely to suffer from diseases. The natural vigor necessary to develop symmetrically is in general a good indicator of strong pathogenic resistance.

The symmetry or asymmetry of a face is sometimes too subtle for people to consciously recognize, but it does influence what both men and women consider attractive. David Perrett of Saint Andrews University (Scotland) used a computer to manipulate pictures of faces, changing the degree of symmetry in each. In this study Perrett's subjects tended to consider the more symmetrical versions to be more attractive.

"A fit, healthy body—that is the best fashion statement."
- Jess C. Scott (Singaporean born American Author)

Besides facial symmetry, what else do men look for in a woman's facial features? Dr. Michael Cunningham, a psychologist at the University of Louisville, Kentucky, investigated that very question. What Dr. Cunningham discovered (Daniel Goleman, "Equation for Beauty Emerges in Studies," *The New York Times,* August 5, 1986) was that elements of the perceived ideal of the attractive female face included the following:

- eye width that is three-tenths the width of the face at the eyes' level
- chin length that is one-fifth the height of the face
- distance from the center of the eye to the bottom of the eyebrow that is one-tenth the height of the face
- height of the visible eyeball that is one-fourteenth the height of the face
- width of the pupil that is one-fourteenth the distance between the cheekbones
- total area for the nose that is less than 5 percent of the area of the face

And apparently, very small differences in these ratios made a large difference for attractiveness. For example, the ideal mouth was found to be 50 percent of the width of the face at mouth level. If that percentage varied by as little as 10 percentage points, the face was found to be much less attractive.

It should be noted that there are some variations in sought-after facial features that vary by race and culture. However, the important takeaway is that there is subtle evolutionary pressure to select healthy, genetically true mates.

Body Symmetry and Beauty

How about body symmetry? In a study (Michael Hopkin, "Dancing Advertises Sexual Quality," *Nature,* December 21, 2005) concerning body symmetry, anthropologists at Rutgers University, collaborating with computer scientists at the University of Washington, describe how they created computer-animated figures that duplicated the movements of 183 Jamaican teenagers dancing to popular music. The figures were gender-neutral, faceless, of equal size, and were evaluated by the researchers for body symmetry. Peers of the dancers were then asked to evaluate the figures as to dancing ability and attractiveness. Symmetrical body figures that danced skillfully were highly rated. This study represents the first time scientists have been able to link body symmetry and skillful dancing to desirability and attractiveness. Rhythm, dancing, and coordination in general are perceived as evidence of genetic quality.

> *"Dancing is a perpendicular expression of a horizontal desire."*
> — *George Bernard Shaw (Anglo-Irish playwright)*

Besides symmetry, *paleo-man* had his sights on other aspects of the female figure. When girls reach puberty their bodies take on certain changes that signal their readiness for mating. For instance, their breasts start to develop and their hips start to widen. This extra fat can act as a beneficial energy reserve during pregnancy. Women who are of childbearing age have waist measurements that are between 67% and 80% of their hip size. Men, children, and older women have waists that are between 80% and 95% of their hip size. The lower waist-to-hip ratio on the part of women in their childbearing years correlates with youth, health and fertility, and our evolutionary inheritance has made us look for and desire these proportionalities.

Devendra Singh, a psychologist at the University of Texas in Austin, surveyed men and women of varying ages and cultures (1990) by showing them photographs of females with varying hip-to-waist ratios. A ratio between 60% and 70% seemed to be considered the most attractive. We should note that this optimum hip-waist ratio held up with both thinner and heavier women. What seemed to matter most was the proper hip-waist proportion.

A Sexy Color

Color plays a role too. In many non-human primate species, including baboons and chimpanzees, females display red on their body (e.g. chest and genitalia) when nearing ovulation. Among humans, sexual excitement is often associated with redness in the body's erogenous areas and facial blushing. Robust physiological processes such as strong blood flow and high testosterone levels (in men) are required to produce a reddish skin appearance. Thus, red may have become a proxy signal for reproductive potential.

To that point, a study (Elliot, Niesta-Kayser, Feltman, "Red and Romantic Behavior in Men Viewing Women," *European Journal of Social Psychology,* July 29, 2010) found that men reported higher sexual attraction toward women dressed in red compared to women dressed in other colors. Men also expressed the intent to spend more money on a date with a woman in red. In a 2012 study (N. Gueguen and C. Jacob, "Clothing Color and Tipping," *Journal of Hospitality and Tourism Research,* August, 2012) involving 272 restaurant customers, researchers found that male patrons gave higher tips to waitresses wearing red compared to waitresses wearing white. And let us not forget that red roses are often associated with sexual intimacy.

A Sexy Smell

Mate selection is not just a visual process; our sense of smell can play a part as well. Odors are known to play an important role in animal mating habits. Animal studies have shown that male testosterone levels are influenced by odor signals emitted by females, particularly when they are ovulating.

Psychological scientists Saul L. Miller and Jon K. Maner, from Florida State University, wanted to see if a similar response occurs in humans. In two studies reported in *Psychological Science* ("Scent of a Women," January 13, 2010), women wore t-shirts for 3 nights during various phases of their menstrual cycles. Male volunteers were asked to smell the t-shirts that had been worn by the female participants. Some of the male volunteers smelled "control t-shirts" that had not been worn by anyone. Saliva samples for testosterone analysis were collected before and after the men smelled the shirts. The results revealed that the men who smelled the t-shirts of ovulating women subsequently had higher levels of testosterone than the men who smelled the t-shirts worn by non-ovulating women (or control shirts). In addition, after smelling the shirts, the men rated the odors on pleasantness. They rated the shirts worn by ovulating women as the most pleasant-smelling.

In a similar experiment (1995, mentioned in a previous chapter, but worth repeating), Swiss zoologist Claus Wedekind had a group of female college students smell t-shirts that had been worn by male students for two nights. The men's t-shirts were void of deodorant, cologne, or scented soaps. Overwhelmingly, the women preferred the odors of men with dissimilar immune system antibodies (called MHCs—major histocompatibility) to their own. A similar study in 2005 showed similar results.

These experiments, and others like them, suggest that odor plays a role in human sexual selectivity, and that evolution has provided us with yet another tool to enhance our reproductive success.

Abnormal, but Not Unnatural

We can use the topic of human sexuality to highlight something critically important about evolution in general, and a *paleo perspective* in particular. It has to do with normal, ubiquitous behaviors as opposed to behaviors exhibited by a small minority.

Homosexual behaviors, practiced by both humans and non-humans, are not the general rule. As mentioned previously, human homosexuality is in the range of 10% or less. Likewise, the vast majority of women do not masturbate regularly, and although males indulge a bit more, it does drop off significantly with age.

Child molestation is another case in point. According to "The Abel and Harlow Child Molestation Prevention Study" (2002), there are estimates that between 1% and 5% of our population molest children. Most would agree that child molestation is a horrible and destructive act, and is not victimless as most acts of consensual homosexuality are. However, relatively few people are sexually attracted to children and this proclivity is therefore not the norm. (However, this fact does not mitigate the horrific nature of this behavior and the damage it does to the young victims.)

Whether it is homosexuality, self-eroticism, or child abuse, we can consider these behaviors to be abnormal in the sense that they are not typical or usual. But that doesn't make these behaviors unnatural. They occur in both human and non-human populations, and with varying frequency. One might wonder why they occur in nature at all. Why didn't evolution and the forces of natural selection eliminate these non-productive, and sometimes destructive, behaviors?

Evolutionary Imperfections

The focus of *The Paleo Perspective* (and evolutionary psychology in general) is to try to account for both physical and behavioral adaptations in the light of our evolutionary past. Why do seemingly non-productive, potentially destructive behaviors continue? The answer to this question lies in the understanding that nature—as a "designer"—need not be a perfectionist. The designs need only to be good enough for survival.

Let us consider several examples of what one might consider to be "design flaws." (This concept was touched upon in Chapter 2.)

Think about the fact that men have nipples. Obviously, men's nipples do not serve the same purpose that women's nipples serve—or any purpose at all. Why then are nipples part of the male design?

The fact is that for the first several weeks a developing embryo follows a "female blueprint." This includes the development of nipples. Only after about 60 days does the hormone testosterone kick in (for males), changing the genetic activity of cells in the genitals and brain. Male nipples, being harmless, do not get weeded out by natural selection because they are not detrimental to human survivability (evolutionary fitness).

Other examples of design "irregularities" include the appendix, the coccyx (tail bone), and "goose bumps." All of these most likely had a purpose in our evolutionary past, but have not been selected out because they are relatively harmless. They do not impact survivability.

Or, consider the fact that as we age we are more susceptible to various maladies such as back pain, joint stiffness, and arthritis. However, back in Paleolithic times many people did not live past thirty. These inherent imperfections in our body design do not generally manifest themselves in the first 30 years of life, and so were not relevant to survivability in Paleolithic times. Having these age-related imperfections in modern times likewise does not affect human reproductive success because most of these maladies affect individuals long past their reproductive years.

> *"Man was made at the end of the week's work, when God was tired."*
> *- Mark Twain (American writer)*

There are also a few behaviors left over from the past that seem to have no discernable modern purpose. The "palmar grasp" reflex is thought to be a vestigial behavior displayed by human infants. When placing a finger to the palm of an infant, the baby will not only grasp it, but the grasp is surprisingly strong. This grasp is also somewhat evident in the feet. When a baby is sitting down, its prehensile feet assume a curled-in posture, similar to that observed in an adult chimp. These reflexes were purposeful in our past when our ancestors' young would have to hold on to the fur of their mothers as they moved about, or perhaps to grasp tree branches.

Another case in point is the hiccup, which is considered to be an evolutionary remnant of earlier amphibian respiration. Amphibians—such as tadpoles—gulp air and water across their gills via a rather simple motor reflex akin to mammalian hiccupping. These behaviors, and many others, have not been selected out because they are not significantly detrimental to survival.

> *"Chaos was the law of nature; Order was the dream of man."*
> *— Henry Adams (American historian and author)*

There is a certain degree of randomness in nature, and man happens to have an instinctual aversion to randomness—and chaos in particular. Try as he may, man cannot control everything, nor can he explain everything. Nevertheless, it is in his nature to try.

Einstein is reported to have said, "God doesn't play dice with the universe." It was in reference to his uneasiness with some of the more arcane and seemingly arbitrary laws of quantum mechanics. His intuition, like most people, was for things to make sense, that there should be a reason for why things are the way they are. Maybe someday, in the very distant future, we will come to understand most everything. But right now it appears that nature can have a mysterious side, an adventitious side, and sometimes even a serendipitous side.

Evolution is a complex process. As we evolved over millions of years we were not re-designed from each earlier ancestor from the "ground up"—over and over again. Mother nature did not redesign today's living things by throwing out the "old model" and start from scratch each and every time. Each subsequent design just needed to be better adapted to survive in the ever-changing environment. It didn't need to be perfect— just good enough. Good enough to survive.

CHAPTER 14

PALEO MORALITY

"Every man must decide whether he will walk in the light of creative altruism or in the darkness of destructive selfishness." – Martin Luther King Jr.

In this chapter we will explore those particular human behaviors that have to do with the concepts of morality, altruism, and ethics. Although there are some differences between the three, there is also a great deal of overlap. Sometimes one of the words is used to explain or define the others. Ethical behavior is also often considered to be moral behavior. Likewise, altruistic behavior is usually considered highly moral and ethical as well.

People who consider themselves to be religious (many of whom believe in a god) might argue that what we consider to be moral or ethical behavior has its genesis in the theological underpinnings of religion. But there is much evidence to suggest that man has an instinctual predisposition to be "good."

By applying a *paleo perspective* we shall see that human morality, altruism, and ethics are indeed evolutionarily based, and are in fact not entirely unique to our species. Let us put the topic of religion, and for that matter the existence of a supreme being, aside for the time being. We will explore those ideas in a later chapter specifically dedicated to the topic of religion. It should also be pointed out that one can believe in a supreme being and still believe in the principles of evolution—they are not mutually exclusive concepts.

Animal Altruism

"Animals have genes for altruism, and those genes have been selected in the evolution of many creatures because of the advantage they confer for the continuing survival of the species." – Lewis Thomas (physician and essayist)

There are examples of altruism in the non-human animal kingdom. Reciprocal food sharing is a form of altruism, where one individual gives up the food it has foraged to another individual. Food sharing has been observed in a wide range of animals including insects, birds, cetaceans, vampire bats, and primates. It is not always an active behavior. Tolerance of theft may also be considered a form of food sharing. Not only does food sharing occur among the members of an animal's immediate family, but it occurs among non-kin individuals as well (Stevens and Gilby, "A Conceptual Framework For Non-Kin Food Sharing," *Animal Behaviour,* 2004).

There are other forms of animal altruism as well. Many wildlife researchers affirm that animal altruism is a well-documented behavior. Those who say animal altruism exists cite examples such as dolphins helping others in need or a leopard caring for a baby baboon (*Eye of the Leopard*, National Geographic documentary, 2006).

Two specific examples of animal altruism were reported in the *Good Nature Travel* blog (February 5, 2013) by Candice Andrews. The first example occurred in 2008 when a bottlenose dolphin came to the rescue of two beached whales in New Zealand and led them into safe waters. Without the dolphin's guidance the whales surely would have died. In another incident in New Zealand a group of swimmers were first surprised when dolphins began circling around them, splashing in the water. The swimmers initially thought the dolphins were displaying aggressive behavior, but it turned out that they were warding off sharks.

In another example, from the aforementioned source, a group of sperm whales swimming off the coast of Portugal took in an adult bottlenose dolphin. While cross-species interactions are not uncommon among terrestrial animals, sperm whales are not known for forging nurturing bonds with other species. For eight days the dolphin traveled, foraged, and played with the adult whales and their calves. Sometimes when the dolphin rubbed its body against the whales they would return the gesture. It is interesting to note that the dolphin had an S-shaped spinal deformity. However, these kinds of altruistic animal behaviors are not the norm.

It is easier to find examples of altruistic behavior in species that are more closely related to our own. According to a study in *The Proceedings of the Royal Society B* (January 2014) involving wild Ugandan chimpanzees, chimps that shared food (either giving or receiving) experienced an increase in the hormone oxytocin. Oxytocin is sometimes referred to as the "love hormone." This hormone leads to stronger social bonds and enhanced cooperative behaviors amongst these primates.

Another example of primate altruism was reported by the *Associated Press* on December 26, 2014:

> *"Onlookers at a train station in northern India watched in awe as a monkey came to the rescue of an injured friend— resuscitating another monkey that had been electrically shocked and knocked unconscious. The injured rhesus monkey had fallen between the tracks, apparently after touching high-tension wires at the train station in the north Indian city of Kanpur. His companion came to the rescue and was captured on camera lifting the friend's motionless body, shaking it, dipping it into a mud puddle, and biting its head and skin—working until the hurt monkey regained consciousness. The first monkey, completely covered in mud, opened its eyes and began moving again."*

While this particular example of altruistic behavior is perhaps rare, it does indicate the capacity for altruistic behavior in monkeys. We must also be cognizant of the fact that scientists do not have the resources or the capacity to record monkey behavior 24 hours a day, 7 days a week. Similar types of altruistic behaviors might not actually be that rare.

Human Altruism

"The purpose of human life is to serve, and to show compassion and the will to help others." – Albert Schweitzer (French-German theologian and physician)

As pointed out in previous chapters, examples of human altruism and moral behavior go back over one million years. Several examples were given in which sick or disabled individuals were cared for. Amongst the examples cited were the Koobi Fora site in Kenya (1.5 million years ago), the Sima de los Huesos site in Spain (500,000 years ago), Shanadar Cave in Iraq (70,000 years ago), and the Dolni Vestonice site in the Czech Republic (25,000 years ago). Caring for the sick is considered an altruistic act, and most people would label this behavior as being moral as well.

Humans evolved in highly social, communal groupings where cooperation, sharing, and reciprocal altruism were commonplace behaviors. Hunters would share the kill with each other, and share it with those who remained in the camp to attend to other matters. Childcare and shelter building were also communal endeavors that required cooperation and reciprocal altruistic behavior.

Are Human Beings Inherently Good From Birth?

Let us turn now to the question of infant morality and altruism. The following excerpts are from a transcript from "The Baby Lab" which aired on CBS ("60 Minutes") on Nov. 18, 2012. Lesley Stahl was the correspondent.

"It's a question people have asked for as long as there have been people: are human beings inherently good? Are we born with a sense of morality or do we arrive blank slates, waiting for the world to teach us right from wrong?

"Psychologist Karen Wynn is the director of the Infant Cognition Center at Yale University. Wynn has arranged a puppet show while infants that range as young as 3 months old watch as the drama unfolds.

"A five-month-old infant named Wesley watches as the puppet in the center struggles to open up a box with a toy inside. A puppet in the yellow shirt comes over and lends a hand. Then the scene repeats itself, but this time the puppet in the blue

shirt comes and slams the box shut. Nice behavior...mean behavior...at least to our eyes. But is that how a 5-month-old sees it, and does he have a preference?

"More than three-fourths of the babies tested reached for the nice puppet. Wynn tried it out on even younger babies, 3-month-olds, who can't control their arms enough to reach. But they can vote with their eyes, since research has shown that even very young babies look longer at things they like. Babies, even at three months, looked towards the nice character and looked hardly at all (much shorter times) towards the unhelpful character. So basically, as young as three months old, we human beings show a preference for nice people over mean people."

Paul Bloom of Yale University and author of *Just Babies: The Origins of Good and Evil* (2013) was involved with the above-mentioned study. Bloom states "Babies are born with a rudimentary sense of justice that allows them to judge the action of others."

Similar findings were reported in an article in *Psychological Science* ("Do Infants Have a Sense of Fairness?" February, 2012). Experimenters Stephanie Sloane, from the University of Illinois, and Renee Baillargeon and David Premack, from the University of Pennsylvania, worked with infants from 19 to 21 months of age. Their experiment focused on getting a fair reward for work performed. Findings from their study indicated that these infants already possessed a sense of fairness.

"Altruism is an instinct for survival that may be in our genes –
an internal force for goodness in everyone that begins with birth."
– James R. Ozinger (Author and Professor of Political Science)

A Genetically Inherited Trait?

We see altruistic and moral behaviors every day as we live out our modern lives. Sometimes we even witness what we would consider to be truly heroic behavior. Psychologist Abigail Marsh of Georgetown

University looked at the brain scans of kidney donors. What she and neuroscientists at Georgetown discovered was that the right side of the brain's amygdala region was slightly larger for these highly altruistic individuals, and conversely, smaller for psychopaths. We also know that our brain has a reward system that involves a chemical called dopamine. Dopamine is a neurotransmitter, and when we do something altruistic or morally correct we are rewarded with a dose of dopamine. Neuroscientists also believe that we have "mirror neurons" in our brain that lead us to feel pain when we see other people suffer.

In the last decade, researchers have found variations of specific genes that can be linked to altruism, cluing them in to its genetic origins. In a recent study published in the journal *Social Cognitive and Affective Neuroscience* (Kristi Eaton, "Is There an Altruism Gene?" *Greater Good Magazine,* January 26, 2011), German researchers discovered that people with a certain variation of the COMT gene (VAL/VAL or VAL/MET variation) donated twice as much money to charity as people with the MET/MET variation of the COMT gene. About 75% of us carry the VAL/VAL or VAL/MET variation, and 25% of us carry the MET/MET variation.

The growing field of "epigenetics" (the study of changes in an organism caused by modifications in gene expression rather than alterations in the genetic code itself) is beginning to provide unexpected insights into the nature of social behaviors like altruism. In studies of genetically identical twins, researchers have found that 30 to 60 percent of altruistic tendencies, such as helping a stranger or donating to charity, can be explained by genetics, with differences influenced by our social or cultural environment (JoAnna Wendel, "Epigenetics Sheds New Light on Altruism," *Genetic Literacy Project,* September 10, 2013). You may recall from a previous chapter that social and environmental factors can affect the expression of genes.

While it seems that humans have a genetic capacity to be empathetic and moral animals, this doesn't mean that we are capable of being highly moral, ethical, and altruistic 24 hours a day, every day. Human behavior isn't that simple. Let us think back to an earlier chapter when alphabet soup was used as an analogy for our genetic inheritance in an effort to help us more easily understand how our genetic makeup interplays with our environment. Most humans have "genetic messages" to behave ethically and morally, but these genetic propensities might be overwhelmed and overcome by external forces and not get expressed. Large group settings, and modernity in general, can disturb one's moral center of gravity. This brings us to the idea of a "duality" in human nature.

Man's Dual Nature

"Individual versus group selection results in a mix of altruism and selfishness, of virtue and sin, among the members of society."
- E. O. Wilson (American biologist, naturalist, and author)

Unless you live on an island all by yourself, thousands of miles from the nearest living soul, you are not just an individual—you are a member of a group. And, since natural selection operates on both the individual and the group level, man demonstrates both selfish and altruistic behaviors.

In an effort to understand and explain human nature, social scientists, philosophers, and theologians sometimes ask us to imagine human beings as having a little angel on one shoulder and a little devil on the other. Each is vying for influence as we go about our daily lives making both small, mundane decisions and important, life-altering decisions. In his first inaugural address, President Abraham Lincoln, in an effort to reach out to a rebellious South, wished that everyone would be "touched by the better angels of our nature." By employing the "angel-devil" imagery one can more easily grasp the idea that each of us is capable of both good and bad behavior.

Selfish versus Selfless

The Paleo Perspective does not look at this inner conflict as a struggle between the invisible forces of "good" and "evil"—if such forces actually exist. It looks at this struggle in evolutionary terms, as the conflict between self-serving behavior and group-serving behavior—"me" versus "us." As pointed out in previous chapters, we all have genetically-based instincts for both selfish (survival of the individual) and selfless (survival of the group) behavior. *The Paleo Perspective* considers moral, altruistic, and ethical behavior to be that which benefits the group at large—many times at the individual's expense. Conversely, immoral and unethical behavior would be that which benefits the individual at the group's expense.

For example, a person may decide to cheat on their income taxes, but that would mean that other people would have to make up the difference in lost tax revenues. Or, if a factory owner saved money by dumping untreated waste into a nearby river, the factory owner would benefit financially, but everybody downstream would suffer. In both cases, self-serving behavior is at the group's expense. These would be examples of unethical, immoral behavior.

But if that factory owner went through the added personal expense of treating the factory waste before it entered the river, the people downstream would benefit. This would be an example of group-serving behavior instead of self-serving behavior. This would be an example of ethical or moral behavior.

Along the same lines, imagine that people are being asked to conserve water during a drought. Some individuals might choose to ignore the plea and use as much water as they desired. But this would be at the expense of their neighbors and the community at large, and would therefore be considered immoral behavior.

Or consider the example of slavery in America during the 17th, 18th and 19th centuries. Individual landowners satisfied their desire for cheap labor and profit by enslaving people of another race. This was a case of individual gratification at the expense of a large number of other people. Slavery precipitated the American Civil War, where perhaps as many as 600,000 people lost their lives—all for the benefit of a selfish few.

Other examples abound. When one gives to a charity, one accepts less for oneself so that others can have more. If someone donates a kidney, they are sacrificing something of value to themselves so that another can benefit. If a soldier sacrifices or endangers his life so others may live, this is altruistic behavior. This is moral behavior. Altruistic and moral behavior is behavior that evolution has programmed into each of us so that the group as a whole benefits.

That being said, one can think of situations where doing what is best for the group also benefits the altruistic individual member of the group.

For instance, by not adding to air pollution, the individual as well as the group benefits with cleaner air. Or, by keeping your own property neat and well groomed, not only do your neighbors benefit (higher real estate value), but you do as well.

The key point here is that we should think of moral, altruistic, and ethical behavior as that which benefits other people—the group. And we should think of immoral, self-centered, and unethical behavior as that which benefits the individual at the expense of the group.

"Morality is the herd-instinct in the individual."
- Friedrich Nietzsche (German philosopher)

Although each of us is capable of both types of behavior, there are times and situations that bring out the best in us. We have all observed that during disaster situations—such as hurricanes and earthquakes—people often respond to the situation with altruism and valor. People risk their lives to help someone else.

In a way, these disasters simulate the urgency and precariousness of the Paleolithic setting, triggering some socially benevolent genetic proclivities (altruism). And so even people who some would consider to be self-centered can act in socially benevolent ways if the situation warrants.

Good and Evil

The Paleo Perspective, being a scientific perspective, does not consider "goodness" or "evil" to be some sort of supernatural quality. However, sometimes behavior is considered so immoral, so unethical, that some people choose to use the word "evil" to describe such behavior. (Mental health professionals might choose to use the term sociopathic or psychopathic instead.) As an example of such behavior one might think of Dr. Josef Mengele the Nazi concentration camp physician. He was known as "The Angel of Death" and engaged in some of the most atrocious medical experiments in history, sending hundreds of thousands to their deaths. Indeed, many people would choose the word "evil" to describe Dr. Mengele's behavior.

Sometimes an individual comes to exemplify what many would

consider to be an exceptionally good person. Mother Teresa might come to mind. Here we have an individual who selflessly dedicated her life to helping those in need. This is in sharp contrast to Mengele inflicting harm on so many. When we think about these two individuals we can easily distinguish good from evil—moral from immoral.

Self-Serving Moral Standards

Man is a socially complex creature that is certainly not capable of being moral 24 hours a day, 7 days a week. He is inherently and concurrently both altruistic and self-serving. He is neither all good nor all bad. "Good" people are capable of doing evil deeds and "evil" people sometimes do moral deeds.

And man, being a complex and clever creature, can sometimes adjust his self-declared moral principles to suit himself. In a recent study ("Moral Judgments Follow the Money," *Proceedings of the Royal Society,* October 29, 2014), Peter DeScioli discovered that people sometimes establish personal moral standards that in fact benefit their own self-interest.

As a case in point, consider the following issue. For many people the acceptance of man's role in global warming has a moral dimension to it. But for people who live in coal or oil producing states (Oklahoma, Texas, etc.), it is hard for them to acknowledge the effect that the burning of fossil fuels has on global warming because their livelihood may depend on a robust petroleum industry. However, the people who live in non-oil producing states (New York, Connecticut, etc.) find it easier to have a moral objection to the continued use of fossil fuels because their livelihood doesn't depend on their continued use.

Consider also the individual who owns or works for a company that sells over-the-counter herbal supplements marketed to improve one's health. There may be no valid and verifiable scientific evidence that these supplements work, or that they are even entirely safe. However, it would be very difficult for the business owner or the employees of that company to take a moral stand against the sale of "useless" herbal supplements. Their livelihoods would suffer.

A quote by American author Upton Sinclair Jr. perhaps sums it up best: "It is difficult to get a man to understand something when his salary depends on his not understanding it." Self-interest can indeed affect one's moral stance on any given issue.

Cooperators and Defectors

Society is a mixture of individuals—some more altruistic than others, some more self-centered than others. Martin A. Nowak, in his book *Super Cooperators: Altruism, Evolution, and Why We Need Each Other to Succeed* (2011), discusses the idea that the relative proportion of *cooperators* (altruistic) and *defectors* (self-serving) in any given population can vary over time. Self-centered individuals will grow in number and will flourish in a highly cooperative society because defectors will take advantage of the cooperative, altruistic majority. If the proportion of defectors becomes too great, the society as a whole will suffer. The survival of the society will become endangered and a crackdown on lawless, self-serving behaviors will therefore become necessary. This crackdown will swing the social pendulum back toward moral, cooperative, and lawful behavior.

Societies and civilizations do not tend to exist in a state of static equilibrium, but rather in a state of dynamic equilibrium. Over time, the proportion of cooperators to defectors for any given society can vacillate back and forth. Sometimes a society can become so non-cooperative and full of defectors that the society (or the civilization) ultimately perishes. History is full of examples of failed societies and collapsed civilizations.

Let us conclude this discussion with the observation that nature has outfitted human beings with a dual nature—altruistic and moral at times, and self-serving and unethical at other times. This is a direct result of evolution's simultaneous selective pressures on both the individual and the group level.

In a previous chapter, the notion of free will was discussed in relation to one's genetic inheritance. The concept of free will also enters into any discussion of morality—or choosing good over evil. And so it is perhaps apropos to consider the following parable, attributed (authorship is in dispute) to the Native American Cherokee tribe:

"The Tale of two Wolves"

A grandfather is talking with his grandson and he says there are two wolves inside of us which are always at war with each other.

One of them is a good wolf, which represents things like kindness, bravery and love. The other is a bad wolf, which represents things like greed, hatred and fear.

The grandson stops and thinks about it for a second, then he looks up at his grandfather and says, "Grandfather, which one wins?"

The grandfather quietly replies, the one you feed.

CHAPTER 15

PALEO RELIGION

"I do not feel obligated to believe that the same God who has endowed us with sense, reason, and intellect has intended us to forgo their use." - Galileo Galilei (16ᵗʰ century Italian astronomer, physicist, engineer, philosopher, and mathematician)

A majority of people would agree that most modern-day religions advocate for what we consider to be moral behavior. *The Paleo Perspective* proposes that human religiosity, just like morality and altruism, is a byproduct of evolutionary forces and is therefore deeply encoded in our DNA.

The Paleo Perspective does not argue against the existence of any kind of supernatural or divine entity. One cannot prove the non-existence of any supernatural deity, just as one cannot prove that former President Barack Obama is *not* a Muslim. (One cannot prove a negative.) Some might suggest that our sense of morality predates our modern concept of God. So let us put aside the supernatural and focus on the natural world. Science only deals with the natural world. A *paleo perspective* is an anthropological perspective, and argues that religions would exist with or without the existence of a god.

Religiosity is in fact a perfectly natural and ubiquitous phenomenon, found in every corner of the globe and on every continent—even on a remote island. There are over 4,000 religions being practiced today, with approximately 84% of the world's people belonging to, or associating with, one of these organized religions (Pew Research Centers Forum on

Public Life, 2010). Additionally, some individuals do not associate with any particular religion, but do believe in some sort of divine or spiritual essence, bringing the total of "believers" up to about 89%. Only about 11% of the world's people are agnostics or self-declared atheists (Gallup International, 2015).

As an example of this persistent religious predilection, consider the communist revolutions in both China and Russia during the last century. Both tried to do away with religious expression and establish atheistic, secular societies.

The former Soviet Union was the first state to have the elimination of religion as an ideological objective. Toward that end, the Communist regime confiscated church property, ridiculed religion, harassed believers, and propagated atheism in the schools. With the collapse of the Soviet Union, and as soon as the antireligious governmental forces were removed or weakened, religious expression gradually returned.

Today Russia recognizes Christianity, Islam, Buddhism, and Judaism as legitimate religious faiths. And although the Chinese Communist Party is officially an atheistic organization, China today recognizes—among others—Buddhism, Taoism, and Confucianism. Governments are working against human nature when they try to eradicate religious expression, and they are almost always unsuccessful when they try.

The Roots of Religion

The Spirit World

If we search for historical evidence of religious behavior, we can find it throughout our past. Archeological findings suggest that human religiosity goes back at least 100,000 years.

A case in point is Skhul Cave in present day Israel, where red ochre and animal parts were found at a human burial site. Anthropologists believe that this kind of ritualistic behavior implies the belief in an afterlife and belief in some sort of animal spirituality. (One theory about the use of red ochre in burials is that it is an attempt to anoint the corpse with some blood-like coloring, symbolic of life.)

More recent examples of spirituality include several caves discovered in France and Spain—Lascaux and Altamira being two of the better-known locations. The paintings in these caves are estimated to be about 15,000 years old, and seem to indicate that these caves were sacred places. Anthropologists interpret these painted scenes as attempts to get in touch with the spirit world in order to ensure successful hunts. We see men wearing animal skins and animal horns, most likely as an attempt to capture the powerful and invisible spirit that they felt resided within the animal. We also see figures that possibly represent shamans. Shamans were individuals (medicine men/women) who possessed special powers, one of which was the ability to cross the barrier into the spirit world.

Additionally, many figurines were found in the areas, which seem to be fertility figures or mother-goddess figures. The possession of a fertility idol is seen as an attempt to curry favor with whatever spirit that may have influence on human fertility—somewhat akin to a lucky or magic charm.

> *"We are not human beings having a spiritual experience.*
> *We are spiritual beings having a human experience."*
> *— Pierre de Cardin (French Jesuit priest and paleontologist)*

Let us look at just a few more examples of early human religiosity, keeping in mind that this chapter is not meant to be an in-depth study of ancient religious practices.

Megaliths and Pyramids

Megaliths ("big stones") were the world's first religious stone monuments. The earliest megaliths were raised almost 7,000 years ago on the Atlantic coast of Europe. Archeologists agree that their prime purpose was the commemoration and veneration of ancestors. At this point in time, many hunter-gatherer cultures had evolved into agricultural and pastoral societies, giving the land beneath them new meaning. Local tribes considered the land to be *their* land—the land of their ancestors. And so these megaliths not only marked out territory, but also were a testament to the presence of ancestral spirits.

Natural or rough-hewn slabs, some weighing many tons, were dragged to a site and were tilted or hoisted into place to form standing stones. These stones, with capstones, formed chambers and immense covered galleries and passages. They were clearly considered to be sacred places where the living attempted to commune with the spirits of honored relatives.

As time went on, civilizations became larger and more complex. Great towers, temples, and tombs were constructed to commemorate the passing of god-like kings. Enormous pyramids served not only as intricate tombs, but also as places where the king-deities began their journey to the afterlife in the spirit world. They too were sacred places. They too were spectacular examples of man's natural affinity for religious behavior.

Some other examples of ancient religiosity include:

- A bull cult at Knossos (Crete, circa 400 BCE)
- The religious burial mounds of Native Americans at Hopewell (North America, circa 450 BCE)
- The pagan religions of the ancient Greeks and Romans (circa 250 BCE)
- The sacrificial burial of the terracotta army of Qin Shihuangdi (China, circa 200 BCE)
- The human sacrificial practices of the Mayan people at Palenque (Mexico, circa 500 CE)
- The Viking queen burial near Oslo (Norway, circa 800 CE)
- The worship of the sun at Machu Picchu (Peru, circa 1400 CE)
- The reading of oracle bones during the Shang dynasty (China, circa 1600 BCE)

While Buddhism, Christianity, Hinduism, Islam, and Judaism are over 1,000 years old, there are religions that began more recently. Scientology, Falun Gong, and Rastafarianism were established in the 20th century and are less than 100 years old. As mentioned earlier, there are several thousand religions being practiced today.

Putting aside any personally held beliefs in the existence or non-existence of "God," let us accept the fact that human beings have been practicing some form of religiosity or spiritualism since the origins of our species. Religious behavior is omnipresent, pervasive, and universal, and is genetically-driven behavior. Furthermore, religious behavior is adaptive behavior, and is a result of evolutionary forces operating at the group level to facilitate cooperation. (You may recall that this idea was discussed in a previous chapter that focused on group unity.) If religiosity is adaptive behavior, what functions does it serve? How does it help us cope and survive as a species? Let us focus on some of the benefits of religion.

The Role of Religion

"Religion is part of the human make-up. It's also part of our culture and intellectual history. Religion was the first attempt at literature, the texts, our first attempt at cosmology, making sense of where we are in the universe, our first attempt at health care, believing in faith healing, our first attempt at philosophy."
- Christopher Hitchens (Anglo-American author, religious and literary critic)

In our earlier discussion on the human brain, it was pointed out that humans need an "operating platform" from which to proceed. We need a worldview narrative as a framework from which all considerations of reality are to be based. A person's religion is an important part of their worldview, and most religions do provide a framework from which to operate. Religious doctrines explain the origins of the universe and all the living creatures found on Earth, including mankind. They also offer a treatise on human nature.

Explaining Creation

Various religions offer an assortment of creation stories. For example, the Book of Genesis (in the Hebrew bible) tells of God's creation of the Earth, the heavens, and all Earth's creatures—including mankind—all done in six days. There is a description of how the first human beings (Adam and Eve) fell from the Creator's grace, and as a consequence, were expelled from paradise (the Garden of Eden). Man became a flawed creature, one that must toil all his life to survive, and then return to the earth at the end of his mortal life. "For you are dirt, and to dirt you shall return." (Genesis 3:19).

The Hindu creation myth says that before time began, there was no heaven, no earth, and no space between. A vast dark ocean washed upon the shores of nothingness and licked the edges of the night. Lord Vishnu slept in these dark waters. Finally, night ended and Vishnu awoke. Out of his navel grew a magnificent lotus flower, and on this flower sat Brahma, the creator. Lord Vishnu commanded Brahma to create the world, including plants and animals of every kind. Some believe that Brahma split in two, creating both human male and female.

For the Hopi (Native American) people, the world at first was endless space, in which only the Creator, Taiowa, existed. This world had no time, no shape, and no life, except in the mind of the Creator. Eventually the infinite Creator created the finite in Sotuknang, whom he called his nephew and whom he created as his agent to establish nine universes. Sotuknang gathered together matter from the endless space to make the nine solid worlds. Then the Creator instructed him to gather together the waters from the endless space and place them on these worlds to make land and sea. When Sotuknang had done that, the Creator instructed him to gather together air to make winds and breezes on these worlds.

There are thousands of creation myths, just as there are thousands of religions. They offer man a creative, imaginative, and easy-to-understand explanation of the cosmos. These religious creation stories are not based on present-day scientific knowledge and do nothing to help explain the extremely complicated workings of our physical universe.

Science only relies on tangible evidence, and so it makes no case for a supernatural and anthropomorphic "Grand Designer." Instead, those with a scientific background might think of nature's randomness as being a kind of creative force, capable of generating boundless possibilities—a "grand designer" of sorts, with no particular design in mind.

Commentary on Human Nature

Most religions endeavor to offer insights into human nature. The Tanakh (Hebrew bible) has many stories of murder, jealousy, and all sorts of examples of human imperfections and sins. In Genesis (the first book of the Tanakh), the first humans, Adam and Eve, commit the first sin by disobeying God, and so from this point on all humans are imperfect creatures. Judaism, Christianity, and Islam all subscribe to the story of Adam and Eve and their fall from grace.

We can, in fact, find commentary regarding human nature in each of the Abrahamic holy books. For example, from the Tanakh: "Surely there is not a righteous man on earth who does good and never sins" (Ecclesiastes 7:20). From the New Testament (Christian bible): "For out of the heart come evil thoughts, murder, adultery, sexual immorality, theft, false witness, slander" (Matthew 15:19). And from the Quran (Islamic holy book): "He has proved a tyrant and a fool" (Surah al-Ahzab 33:72).

In similar fashion we can find commentary on human nature in the Tripitaka (Buddhist scripture): "Should you find a wise critic to point out your faults, follow him as you would a guide to a hidden treasure." And from the Guru Granth Sahib (Sikh scripture): "Again and again, we hear and tell stories; we read and write and understand loads of knowledge, but still, desires increase day and night, and the disease of egotism fills us with corruption."

But even with all of man's supposed faults, many religions bestow upon him an elevated status. In the Tanakh we find that "God created man in his image, in the divine image he created him; male and female he created them" (Genesis 1:27). God blessed them, saying: "Be fertile and multiply: fill the Earth and subdue it" (Genesis 1: 28). And from the New Testament: "As is the earthy one, so also are those who are earthy; and as is the heavenly one, so also are those who are heavenly. Just as we have borne the image of the earthy, we will also bear the image of the heavenly one" (Corinthians 15:48).

These passages suggest that both Jews and Christians consider man to be special amongst all other living creatures. Likewise, Muslims also consider humans to be special, created with a free will to obey and serve Allah.

It is reasonably safe to say that most religions encourage man to aspire to righteousness, while acknowledging his shortcomings.

Explaining the Unexplained

Religion also attempts to explain various happenings and natural disasters. People in the past felt that events such as earthquakes, tornados, or floods were the work of God, and were quite possibly retribution for bad behavior.

There is the story of Noah in Genesis. In this story, God is so angry at man's wickedness and sinful behavior that he causes a great flood that kills every human on Earth except for Noah and his family. Similarly, in Greek mythology, the god Poseidon was the cause of earthquakes. When he was in a bad mood, he would strike the ground with his trident, causing earthquakes and other calamities.

There are still remnants of this kind of thinking today. Not too long ago some religiously conservative thinkers in the United States opined that getting AIDS was God's retribution for sinful, homosexual behavior. Indeed, it is not uncommon for people to sometimes wonder why they have been singled out and inflicted with a serious disease. They question if they have been punished for some transgression. And even though people feel dismayed by life's seemingly senseless tragedies, many are comforted with the thought that even if misfortune is beyond comprehension, it is still all part of God's plan. It would seem that this type of thinking offers some a degree of comfort and closure, and is preferable to being a victim of nature's indifference.

Divine Intervention

Many military battles and wars have been looked at as examples of divine intervention. For example, the ancient Israelites were victorious over a king named Chedorlaomer, and God is credited with lending a helping hand: "And blessed be God Most High, who delivered your foes into your hand" (Genesis 14:20). In another instance, the Israelites had lost a battle at Ai, near Jericho. And the Lord said to Joshua, "Stand up. Why are you lying prostrate? Israel has sinned: they have violated the covenant which I enjoined on them" (Joshua 7:10).

Religious texts of other faiths also contain examples of people attributing military victories or defeats to the hand of some deity. One such example is the successful Battle of Badr, fought in 624 CE, in the Hejaz region of western Arabia (present-day Saudi Arabia). This was a key battle in the early days of Islam, and a turning point in Muhammad's struggle with his opponents among the Quraish in Mecca. The battle has been passed down in Islamic history as a decisive victory, attributed by believers to divine intervention.

A more recent example of supposed divine intervention can be found in the annals of the American Revolutionary War. On August 27, 1776, just weeks after the signing of the Declaration of Independence, the British forces had George Washington and his Continental Army "on the ropes" in New York City. There seemed to be no possible escape across the East River. As the British closed in, it looked very much like the American Revolution would be over in just a few weeks. However, a dense fog "miraculously" settled in and provided Washington and his ships with the cover they needed to successfully sneak all 9,000 soldiers across the East River.

A Rationalization for Hatred and Violence

Religious doctrine can be used as a rationale for prejudice and hatred toward people of different faiths or different cultures. For example, there was the expulsion of Jews from Spain in 1492 by the Catholic monarchy and the anti-Jewish pogroms in the Russian Empire in the early 20th century.

A more recent illustration of the juxtaposition of God and conflict can be exemplified by various present-day so-called Islamic jihads. These military campaigns usually involve hostilities between Sunni and Shiite Muslims, or conflicts between Muslim fundamentalists and various Western powers. In most of the conflicts the Islamic forces say they are waging a holy war against infidels, and therefore God is on their side. Taken to the extreme, individuals or groups can seemingly justify any level of violence and mayhem in the guise of religious and moral rectitude, leaving them free of guilt and shame.

Most would agree that religion is misused when people claim they are carrying out "God's wishes" in an effort to cover up animosity or to satisfy their own secular desires for power and wealth.

Providing Comfort

Disorder, confusion, and turmoil can precipitate high levels of anxiety in just about anybody. Humans abhor chaos. Pointless randomness is anathema to the human psyche, and we will choose almost anything that will give us a sense of order. Historically, people have even chosen brutal dictatorships over pandemonium. They do so because order is comforting, and everybody understands that anarchy can be very dangerous.

Religion is a very effective antidote to randomness, chaos, and disorder because most religions offer an omnipotent and omnipresent God—a God who is in charge of everything. As previously pointed out, it is comforting to think that God has a plan for each of us, and that the violence and turmoil in the world (as bad as it seems) is still part of a plan—a divine plan. We also know that man operates most effectively in group formation, and groups need good leaders. It so happens that God is the ultimate leader—infinitely wise, all-powerful, and perfectly just. The world is often a chaotic and scary place. It is comforting to believe that something or someone—all-knowing and all-powerful—is fully in charge of everyone and everything.

A study appearing in the *American Journal of Epidemiology*, done by researchers at the London School of Economics (LSE) and Erasmus University Medical Center in the Netherlands, found that the secret to sustained happiness might lie in participating in religion. Mauricio Avendano, an epidemiologist at LSE and one of the authors of the study said "The church appears to play a very important social role in keeping depression at bay and also as a coping mechanism during periods of illness in later life." (Sarah Pulliam Bailey, "Want Sustained Happiness? – Consider Religion," *The Washington Post*, August 18, 2015).

Humans are also anxious about what the future will bring, and have at various times in their history (including today) employed various techniques in an attempt to predict the future. Generally speaking, these techniques involve the expectation that some deity or spirit will influence the outcome of what most would consider to be a random event. There are many of these techniques and rituals. A few examples are:

- Reading tea leaves or cracks on a turtle shell
- Reading the dispersal patterns when casting stones, beans, or bones
- Examining the liver or blood patterns of slaughtered animals
- Reading the patterns of molten lead when thrown into cold water

It is certainly comforting to believe that a benevolent and omnipotent super-being is in charge of the future—including your afterlife.

With regard to human mortality, the idea of dying can produce great anxiety in most people. Fear of death and insignificance is perhaps the byproduct of acquiring a bigger brain and greater self-awareness. The idea of a person, or a person's "soul," existing in perpetuity in some sort of a heaven or paradise is a very comforting thought for most people. The world's two most popular religions, Christianity and Islam, include the concept of an afterlife. Hinduism is the world's third most popular religion, and although it doesn't focus on an afterlife per se, Hinduism does delve into the concept of reincarnation.

Whether it is an afterlife or a reincarnation, it is reassuring to think the soul goes on. Indeed, as one might expect, religions that offer a "death-defying" theology have broad appeal.

Keeping Order

Another very important function of religion is to help keep order in society. Many religions are organized around some sort of a hierarchy. Presumably, God would be at the top of such a hierarchy, with religious leaders (popes, ayatollahs, chief rabbi, etc.) and clerics (priests, ministers, rabbis, imams, etc.) occupying a higher level than the general public. Humans desire a fair chieftain, but tough enough to "clean house" when necessary. Humans do crave leadership and moral authority. Organized religions, with God at the top, can provide order and stability to a community.

Most religions have a set of commandments or rules by which adherents are encouraged to live by. These guidelines are stated in somewhat different forms, depending on the particular religion, but there are many similarities in these guidelines across religious sects. If we took a brief survey of various religious tenets, we would see that most of them would include prohibitions on killing, stealing, lying, committing adultery, and coveting other people's possessions. The Old Testament (Tanakh) also includes suggestions about honoring your father and mother. In general, most religious guidelines encourage altruism and moral behavior, and discourage antisocial behavior. If everybody followed these rules society would indeed be a very safe and orderly place. In such an ideal society, there would be very little need for a police force, or a military for that matter. Utopia!

What happens if religious subscribers fail to uphold the various rules and commandments laid out by their religion? Many religions employ the concept of "sin." By sinning, one risks dishonor, disrepute, or shame. Worrying about one's reputation, feeling shamed, or shunned are powerful social sentiments that help control errant human behavior. Man's inherent capacity to experience guilt, humiliation, and dishonor are useful and adaptive sentiments that help maintain social order.

Most religions also subscribe to some sort of ethic of reciprocity—sometimes referred to as "The Golden Rule." Several examples include:

- "One should treat others as one would like others to treat oneself." (Confucianism)
- "Wish for your brother, what you wish for yourself." (Islam)
- "What is hateful to you, do not do to your fellow." (Judaism)
- "Do to others what you want them to do to you." (Christianity)
- "Hurt not others with that which pains yourself." (Buddhism)

The obvious social benefit of the *Golden Rule* is that it encourages altruism and ethical behavior, and discourages antisocial behavior and conflict.

Connected to the concept of ethical reciprocity and social justice is the idea of an afterlife. The concept of a heaven and a hell encourages good behavior and discourages sinning. One must earn one's place in "paradise" by following the tenets of one's religion, and lead a moral life. We can perhaps see hints of the fear of punishment when we hear expressions such as "God-fearing person," or "put the fear of God in him." One could argue that God is the ultimate enforcer of the rules—the ultimate force for social order and justice.

Although Hinduism does not subscribe to the concept of an afterlife, there is a belief in reincarnation. In the case of reincarnation, the status of one's next life depends on the moral quality of your present life. And so one is encouraged to lead a moral life.

Whether there is a belief in an afterlife or in reincarnation, people not only have incentive to lead a proper life, there is an expectation that those who don't will receive just punishment. Belief in an afterlife (or reincarnation) can be a very effective means to help maintain social order.

Humans, and some of our closely related primates, do have a natural sense of justice and fairness. Studies by Frans de Wall, director of the Yerkes National Primate Research Center at Emory University, suggest a long evolutionary history to our sense of fairness. One such study published (January 2013) in the journal *Proceedings of the National Academy of Sciences* suggests that chimpanzees may show some of the same sensibility about fairness as humans do. Let us also recall those studies done by psychologist Karen Wynn at the Infant Cognition Center (Yale University) mentioned in the previous chapter on altruism and morality. In those studies, human babies (under a year old) demonstrated a sense of fairness and morality. As stated previously, morality and a sense of justice have been encoded in our DNA for a very long time.

Religion satisfies our desire for justice by establishing a moral code to live by, and promising punishment to those who violate those rules. All this helps maintain order in society.

Providing Vital Rituals

Formal religions promote social order in another important way. The practice of most religions usually involves participation in various rituals. These rituals might include initiations, rites of passage, ceremonial events, and regular prayer. Catholics are encouraged to attend mass on a weekly basis. Muslims are expected to pray five times each day. Jewish men are encouraged to wear a yarmulke. There are thousands of religions with thousands of rituals.

While religious rituals might seem pointless to the non-believer, they provide a very important benefit to their practitioners. All rituals, customs, and practices serve to help bond group members together. People who attend a Christian mass pray together and observe the same religious holidays together. Many religious communities have rules about food and dress. Jews and Muslims are both encouraged to not consume pork. And many conservative Muslim sects require their women to wear certain body coverings (burka) and head-scarves (hijab). These shared rituals and customs are all bonding experiences.

There is also evidence that the particular theological underpinnings of a religion are not critical for subscribers to benefit from their religious practice. Consider an article appearing in *The New York Times* (May 29, 2013) written by columnist T.M. Luhrmann. (T.M. Luhrmann is also a professor of anthropology at Stanford University and the author of *When God Talks Back: Understanding the American Evangelical Relationship With God*.) In the article, Luhrmann states that after spending much time in Evangelical churches, he has come to the conclusion that the focus of Evangelical gatherings is not so much on theology or the intellectual question of God's existence but on "how to feel God's love and how to be more aware of God's presence." The Evangelical service promotes a feeling of joy and a sense of God's goodness.

T.M. Luhrmann also makes reference to a French sociologist, social psychologist, and philosopher named Emil Durkheim (1858 – 1917). In his book *The Elementary Form of the Religious Life*, Durkheim avoids focusing on the question of God's existence and instead looks at religion as a social institution. He suggests that religion allows people to experience themselves as members of a group, as part of something larger and more important than their individual selves. As such, religious affiliation serves as a source of collective consciousness, and enhances group camaraderie and solidarity.

Ancestor veneration and worship, for example, serves to connect present group members with past group members, enhancing group identity, and providing a measure of continuity and stability for the group. Religious rituals provide a vehicle through which we can exercise and satisfy our need for the "group experience." The mere act of gathering is an instinctively rewarding experience for us humans, satisfying our Paleolithic need to be part of a group.

In a 2010 Pew Research Center survey of religious and biblical knowledge, the average American only answered half the questions correctly. (It is perhaps somewhat interesting that atheists and agnostics scored the highest.) But theological expertise is not a prerequisite for religious benefits. The *Golden Rule*, various religious commandments, the concept of sin and punishment, and religious gatherings and rituals all have many social benefits. At the very least, religions ask us to conform, and conformity in and of itself promotes order. All of these practices help facilitate group integrity, coherence, cooperation, and social order. If you are a believer in God, then perhaps you think of religion as a gift from God, bestowing upon man numerous social benefits. Believers can also think of the social benefits of evolution as inspired and directed by God.

Enhancing Solidarity Amongst All Believers

As pointed out in a previous chapter, the existence of subgroups within the larger society can precipitate some inter-group conflict. However, often individuals who are ardent practitioners of one particular faith feel they have something in common with the passionate practitioners of other faiths, even though different religions have vastly different practices. Religious people of all faiths share their enthusiasm and commitment to God, and therefore feel a certain bond with each other. And so commitment to religious principles and practices as such can be a binding force for a religiously heterogeneous society. Indeed, simply believing in any sort of deity can be a bonding agent for the believers of all faiths.

According to a 2011 *Gallup Poll*, 92% of the people in the United States believe in God. And although God is not mentioned explicitly in the U.S. Constitution, it is in all 50 state constitutions (Pew Research, 2016). In fact, belief in God can sometimes be required by civil law. There are seven states within the United States whose state constitutions include language that could be used to prevent an atheist from holding public office or performing other civic duties. Three examples are:

Arkansas: Article 19, Section 1

> *"No person who denies the being of a God shall hold any office in the civil departments of this State, nor be competent to testify as a witness in any Court."*

Mississippi: Article 14, Section 265

> *"No person who denies the existence of a Supreme Being shall hold any office in this state."*

Tennessee: Article 9, Section 2

> *"No person who denies the being of God, or a future state of rewards and punishments, shall hold any office in the civil department of this state."*

Just like other customs and traditions, religious affiliation can help people feel connected with the other members of their tribe, country, or society in general. And belief in a supreme deity helps satisfy man's need to group affiliate under a worthy leader. After all, who is a more worthy leader than God?

Homo Religiousus

> *"For those who believe, no proof is necessary. For those who don't believe, no proof is possible." - Stuart Chase (American economist, social theorist, and writer)*

We all would like to believe that human beings are intelligent, rational creatures. We are, after all, at the very top of the food chain, and according to Genesis, "have dominion over the Earth." While we are indeed rational some of the time, there are many times when we are not. People believe all sorts of things. The problem with "belief" is that it need not necessarily be based on fact. One might wonder if our ability to be in a state of belief (or disbelief)—contrary to reality—is peculiar to Homo sapiens.

In his book *The Believing Brain*, author and skeptic Michael Schermer argues that most of our opinions and beliefs, regarding everything from your favorite sports team to politics and religion, are not based on a thorough examination of relevant facts or pertinent information. Instead he says:

"We form our beliefs for a variety of subjective, personal, emotional, and psychological reasons in the context of environments created by family, friends, colleagues, culture, and society at large; after forming our beliefs we then defend, justify, and rationalize them with a host of intellectual reasons, cogent arguments, and rational explanations. Beliefs come first, explanations for beliefs follow."

Superstitions are a good example of irrational beliefs. In general, they are a reflection of our brain's effort to assign causality—even if none exists. Consider the following list of some very common superstitions:

- Friday the thirteenth is an unlucky day
- A rabbit's foot brings good luck
- Finding a four-leaf clover will bring good luck
- If you walk under a ladder you will have bad luck
- If a black cat crosses your path you will have bad luck
- Breaking a mirror will bring you seven years of bad luck
- Opening an umbrella in the house will bring bad luck
- Finding a horseshoe will bring good luck

According to a United Kingdom survey (2004) performed by Professor Richard Wiseman (of the Psychology Department at the University of Hertfordshire), 77% of respondents indicated that they were at least a "little superstitious," and 42% indicated that they were "somewhat or very superstitious." According to a 1996 Gallup survey done in the United States, 53% of those responding reported being a "little superstitious," and 25% indicated that they were "somewhat to very superstitious." And when it comes to belief in ghosts, a 2013 Harris Poll (online survey of 2250 people) indicated that 42% of Americans believed in ghosts. Similarly, an article in the *Huffington Post* (Lee Speigel, "Spooky Number of Americans Believe in Ghosts," February 02, 2013) reported that 45% of Americans believe in ghosts, or that the spirits of dead people can come back in certain places and situations.

Being superstitious, or believing in luck is an outgrowth of the belief in the existence of various supernatural forces at work in the universe. These invisible forces come into play if, for example, you find a four-leaf clover, break a mirror, or carry a rabbit's foot.

Being superstitious or believing in supernatural entities is pertinent to our discussion of religion in that the belief in luck or supernatural forces is not possible without a belief in the existence of some sort of invisible spirit world. Most religions have a god (or several gods)—a god that is spiritual and supernatural. You may recall that belief in ancestral spirits and quasi-religious and ritualistic behavior goes back to the beginning of our species. We are not saying that being religious and being superstitious are exactly the same. What is being suggested is that like superstitions, religions ask us to believe in something that we cannot see or measure scientifically.

As stated earlier, one cannot disprove the existence of supernatural phenomena because they are, by definition, "supernatural." The point is that whether God exists or not, humans are capable of forming beliefs for which they have no logical or scientific basis. As Schermer postulates in *The Believing Brain*, man forms beliefs first, and then seeks to find a rationalization for those already formed beliefs. And those beliefs are not the result of a rigorous analysis of the pertinent facts in the first place. Religions ask us to believe, not analyze.

It should be pointed out that having a set of beliefs, even if those beliefs are not entirely based on fact, can be helpful and beneficial. For example, the New York Mets (an American baseball team) have a slogan: "You gotta believe." This slogan is helpful because it encourages the players and fans to think positively. Thinking positively can not only have a direct, beneficial effect on an athlete's level of play, but fan enthusiasm can also help encourage the players on the field as well. On the other hand, believing that your team consists of a bunch of losers can have a negative effect.

In general, having "faith" can be advantageous in that it allows the individual to create a positive worldview narrative to operate from. One could certainly argue that having religious faith is uniquely human.

The Omnipresent Religious Urge

One also needs to consider the fact that even avowed atheists could have religious urges. A feeling of spirituality might wash over an atheist while in a church, mosque, or temple. Or one might feel a "divine" presence at a religious ceremony or some other somber event. Agnostics and atheists from time to time cry out "Jesus Christ," "oh God," or "Jesus, Mary and Joseph" even though they are non-believers. Perhaps the fact that both believers and non-believers sometimes evoke "the Lord's name" suggests that it is quite natural for humans to seemingly sense the presence of a supernatural entity. The analytic part of one's brain may question the existence of a God, but our religious instincts are real and palpable.

Is There a God Gene?

If religious behavior is genetically induced, is there such a thing as a "god gene"? Dr. Dean Hamer is a preeminent geneticist and the author of several books on genetically-driven behavior. He has appeared on national news shows and documentaries for HBO, PBS, and the Discovery Channel. In his book *The God Gene: How Faith is Hardwired into our Genes* (2004), he describes the elaborate and detailed research that supports the theory of genetically-based spirituality. Hamer argues that spirituality is one of our basic human inheritances. It is in fact an instinct. He also points out that there isn't just one gene involved, but several. Dr. Hamer goes on to say:

> *"The specific gene I have identified is by no means the entire story behind spirituality. It plays only a small, if key, role; many other genes and environmental factors also are involved. Nevertheless, the gene is important because it points out the mechanism by which spirituality is manifested in the brain."*

A God Called DNA

Religion is a powerful narrative, and is often an integral part of an individual's worldview narrative. The stories of Jesus, Muhammad, Abraham, Buddha, and Zarathustra do have an historical basis to them, but they also represent something much deeper within the human psyche. These narratives reflect our sense of right and wrong, our sense of fairness, and our sense of morality. And, as discussed in the previous chapter, these sensibilities are the result of evolutionary pressures to ensure the altruistic and cooperative behavior that is so essential for group functioning.

The abovementioned "prophets" are the personifications of morality, altruism, and cooperation—the Golden Rule if you will. Their messages and the stories that surround their lives are encoded in the strands of our DNA. Organized religions merely formalize and standardize our innate sense of right and wrong, good and bad. The religions that resulted from the lives of the abovementioned prophets are the canonization of our genetic inheritance—the expression of our "moral compass."

Life can be extremely difficult at times. Human capacity for self-awareness in and of itself can be burdensome. Religions are powerful and enticing narratives that offer us explanations for life's mysteries and comfort for our soul. They offer us some relief from the dread of death. For many, they provide a worldview narrative from which to lead their lives. Many religions have come and gone. The ones that have lasted the longest, the ones that have survived the test of time, are those that offer us a life of order, a promise of justice—and above all, hope.

Religious behavior is not a result of happenstance or serendipity, but is in fact evolutionarily driven adaptive behavior. To that point, Jesse Bering, in his book *The Belief Instinct* (2011), argues that the instinct to believe in God and other unknowable forces may have actually given early humans an evolutionary advantage.

We are not perfectly logical creatures such as the one portrayed by Spock in the Star Trek series. Evolution did not demand scientific analysis and uncompromising logic from us. We evolved in a dangerous world—one in which our instincts and beliefs were critical for our survival. Ancient man did not question or analyze his instincts, nor did he agonize over the accuracy of his beliefs. His opinions and beliefs were based on whatever worked to ensure his survival. Our instinct to be religious is a case in point. Perhaps we should consider ourselves to be "Homo religiosus" instead of Homo sapiens.

"If one wishes to form a true estimate of the full grandeur of religion,
one must keep in mind what it undertakes to do for men. It gives
them information about the source and origin of the universe, it
assures them of protection and final happiness amid the changing
vicissitudes of life, and it guides their thoughts and motions by means
of precepts which are backed by the whole force of its authority."
- Sigmund Freud (Austrian neurologist and founder of psychoanalysis)

CHAPTER 16

PALEO ECONOMICS

*"An economist is an expert who will know tomorrow why
the things he predicted yesterday didn't happen today."
Laurence J. Peter (Canadian educator and author, best
known for his formulation of "The Peter Principle")*

Economics is a terribly complicated subject, and this chapter is not
meant to be technical or thorough in its discussion of economics. It is
an attempt to look at some economic topics and concepts such as gross
domestic product (GDP), wealth, unemployment, and capitalism with our
evolutionary past in mind. More specifically, we shall compare economic
life in the Paleolithic past with economic life in modern times. (It will
become a little clearer as this chapter unfolds, but to understand what is
being presented, one needs to consider "modern times" to be approximately
from 12,000 years ago to the present.)

Previously mentioned is the fact that a *paleo perspective* is a reductionist
perspective. In an effort to simplify and clarify some economic concepts,
several hypothetical situations will be proposed. Hopefully, these
hypotheticals will help the reader focus on the essence of a particular
economic concept by stripping away the veneer of modern presumptions.
Looking back at simpler times is a good place to start.

The Paleo Economy

Let us begin by focusing on our hunter-gatherer past. We should recall from an earlier chapter on human evolution that man's ancestors were hunter-gatherers (sometimes referred to as foragers) for several million years. It is only for approximately the last 12,000 years that agriculture, animal husbandry, and manufacturing have become the norm. What was the Paleolithic hunter-gatherer economy like? Can we talk about things like employment, wealth, and gross domestic product (GDP) in such a primitive society?

Paleo Employment

Let us start with the concept of employment. "Paleo man's" occupation (or job) was hunting, food-gathering, and shelter-building. Everybody was employed in the sense that each individual had the need and opportunity to go out and hunt, gather edible plants, and secure shelter. One didn't need to be hired as a hunter or gatherer. Everybody looked to procure food—because if they didn't, they would go hungry. Everybody fabricated or secured some type of shelter—because if they didn't, they would fall victim to the elements. One didn't get "hired" by a third party to secure shelter. And so the unemployment rate was essentially zero, and competition for "jobs" was virtually non-existent.

There might have been some competition for life's necessities (natural resources) between different prehistoric clans or troupes. But for much of man's ancient history, his survival depended more on his cooperation with fellow tribe members than competition with other groups. Ideas were shared. Resources were shared. Life was tough enough, so hunting and food gathering were cooperative enterprises. Job competition and career searches were not an issue.

Paleo Work Skills

We are describing a rather simple economy, but one in which the main character, primitive man, is himself not so simple. In "paleo times" each individual needed to have a working knowledge of his surrounding environment. Rudimentary knowledge of local geography, the weather, the seasons, the sun, and stars were a must. *Paleo man* had a working understanding of the plants and animals in his niche. He needed to possess appropriate hunting, fishing, and foraging skills.

And so primitive man possessed a wide variety of skills and knowledge. He shared this information with other group members, and passed it along to the next generation of clansmen. You might consider early man to be a primitive form of a "Renaissance man," displaying a plethora of expertise—a virtual one-man show. Life was tough. Life was complicated. It required ancient man to be highly skilled, quite knowledgeable, and at the top of his game, both physically and mentally.

Paleo Wealth

What about the accumulation of wealth? *Paleo man* moved around a lot, carrying his possessions with him. How much could he possibly carry? How much could he possibly own? Perhaps he carried a stone hand axe, a spear, an animal skin for warmth, and possibly a glowing ember to start the next fire. There really wasn't much of an opportunity or a benefit to accumulate lots of material goods. Everybody in each generation needed to know how to fashion stone tools and weapons. Each generation needed to know how to skin an animal and fashion clothing. Each generation needed to know how to start a fire. Information and skills were passed on more than material goods were.

What about housing as an indicator of wealth? Depending on how far back we look, we would be talking about cave dwellings or possibly some lean-to type of fabrications. These caves or shelter fabrications would be multiple family dwellings housing the entire band. An individual would have no need of more than one shelter at any given time. Nobody accumulated wealth by "owning" multiple dwellings.

Once we acknowledge *paleo man's* hunter-gatherer, nomadic (or semi-nomadic) existence, it follows that we shouldn't expect to discover much difference in wealth between one individual and the next, or between one family and the next. There were no monopolies, whereby a single individual possessed all the bows, arrows, spears, and stone axes in the troupe. Man lived in cooperative groups, where shelter and food were shared. Consequently, there was little opportunity for wealth inequality.

Paleo man existed on a fairly level playing field. He started with very little and accumulated very little. There were no class struggles. It wasn't until humans established villages, towns, and cities that there was much of an opportunity to accumulate material goods, and consequently the possibility for significant wealth inequality within a society.

Paleo GDP

GDP is the sum of all goods and services in an economy. What would be the sum total of all the goods and services in a "paleo economy"? In part it would be the sum total of all stone tools, weapons, animal skins, shelters (found or built), animals killed, and fruits and berries gathered. In addition, we would also consider "services" such as hunting, tool fabrication, clothing fabrication, shelter fabrication, fire starting and maintenance. Given the nomadic and socially cooperative nature of ancient hunter-gatherer society, the GDP would be relatively small, and fairly evenly shared.

Granted, this is all a very simplistic and reductionist perspective on ancient economics. But it allows us to become more mindful and critical of how different the economic landscape is in modern society. When we think about what comes naturally to us, we must keep in mind that our genetic profile was not formed in modern times, but long ago in Paleolithic times.

Paleo Man in Modern Economic Times

Let us move forward in time to roughly 10,000 to 12,000 years ago, to a point in time that is usually referred to as the beginning of the "Holocene" (Holocene meaning "recent"). The last ice age had subsided and the climate on Earth began to warm. Not only did the average temperature on Earth increase, but the Holocene also brought with it a period of climatic stability, with less severe swings in temperature. (As previously stated, let us consider this time period—from roughly 12,000 years ago to the present—to be "modern times.") These relatively mild temperatures and stable climate patterns were kind to most fauna and flora on the planet, including Homo sapiens.

With the climate now being more cooperative, humans were inspired to take advantage of the improved conditions and eventually figured out how to domesticate both plants and animals. The domestication of plants and animals allowed man to become less nomadic, enabling him to settle down on the surrounding land that he was in the process of domesticating. Little villages grew into towns, towns grew into cities, and cities grew into states. This significant change in life-style is often referred to as the "Agricultural Revolution" or the "Neolithic Revolution."

We should note, as a species becomes more successful and more densely configured, resources within its eco-niche could become stressed. The resultant competition within the species for these resources is sometimes referred to as "intra-species conflict."

The Creation of a Market

Humans now lived in closer proximity to each other—close enough to prompt and facilitate the trading of goods and services. Trade allows one to specialize and exchange one's specialty for the specialties that others have to offer. For example, one could raise and exchange one's goats for corn or some other crop. One could possess carpentry skills and exchange those skills for meat or crops. This exchange of goods and services is a type of "market." Nobody had to invent this market. It came about quite naturally as man transitioned from a nomadic hunter-gatherer lifestyle to a more settled one based on animal husbandry and the domestication of crops.

This drastically changed how people went about surviving—how they went about "making a living." The economy evolved from the previously described *paleo economy* to a market economy characterized by specialization, diversification, and the exchange of goods and services.

It should be pointed out at this point in the chapter that no distinctions are being made between capitalistic, socialistic, or communistic economies. For the sake of clarity and simplicity, the term "market" simply refers to the necessary exchange of goods and services in a diverse and specialized economy. (It also doesn't refer to the "stock market.")

The Exchange of Goods and Services

As a result of the Neolithic Revolution, individuals only needed to specialize in one or two skills. This is especially true in very recent times, where most of us do not have to be a "jack of all trades" like primitive man was. An individual can focus on being a carpenter, electrician, doctor, accountant, farmer, psychologist, etc.

For example, a doctor need not know how to build a house, grow food, or do tax returns. A doctor can get by with very little knowledge of geology, or the animals and plants that are part of his or her ecosystem. Similarly, an electrician need not have expertise in plumbing, farming, or the latest advancements in medical science.

And so a psychologist, for example, can exchange his or her particular set of skills to get a house built, to secure food, and to obtain clothing. One can barter to obtain all these different products and services, but the use of some sort of currency —beads, shells, gold or silver coins, paper bills, bitcoins, credit cards, etc.—makes the whole exchange process much easier. In fact, it is hard to imagine the exchange of goods and services in a large, complex society without the use of some sort of currency or "money."

Nowadays, most people do not have a choice about being part of a market system. All 7 billion of the Earth's current inhabitants could not return to a hunter-gatherer system. The vast majority of the world's people now live in relatively densely configured villages, towns, and cities. As such, they do not have access to large stretches of wilderness for hunting and food gathering.

Instead, we depend on poultry farms, cattle ranches, and commercial fisheries to satisfy our protein needs. We depend on large commercial farms to grow corn, wheat, barley, soy, and other grains. Indeed, for all practical purposes—as individuals—we lack the opportunity and the ability to feed ourselves. Unless you are living on a remote island, isolated tropical jungle, or on the Arctic Circle, you are dependent on some sort of a market for your food.

Modern man is in a similar predicament with respect to shelter (housing). Even if we possessed the necessary skills to build our own house, as individuals we would still lack the capacity to mill our own lumber, manufacture our own nails, bake our own bricks, and fabricate electrical wiring. We would need to participate in the market to secure those items. It is true, but sad to say, that if an individual (for whatever reason) could not participate in the market, he or she could come to be homeless, even if that individual was highly motivated to fabricate a home for him or herself.

And so we can see that because most humans now live in relatively populated areas compared to the ancient hunter-gatherer, they are forced to participate in a market economy. And in this market, each person needs to offer some skill or some product as barter for all the other things he or she needs or desires. Simply put, that individual needs to have a job.

The Efficiency of Specialization

As previously stated, the market system allows for the diversification and specialization of skills. Diversification and specialization can result in huge improvements in the way society provides food, shelter, clothing, and all the other things that are essential for survival. One should recall that in the *paleo economy*, most everybody needed to be sufficiently good at just about every task that they were confronted with—hunting, tool making, shelter building, etc. But if everybody is only required to contribute the skill that they are most proficient at, then the whole process becomes much more efficient.

For example, someone who is exceptionally good at hunting would just hunt for the community. Someone who is exceptionally talented at shelter fabrication would just do construction for everybody. Somebody who was a good cook would just cook for the group. Each of these people would market their skills in return for the other goods and services that they needed. Since everybody was specializing in the task they were most talented, theoretically, the whole group would benefit.

Of course this kind of specialization is only feasible in larger groupings—the kind of groupings that the Neolithic Revolution brought about. It is not likely that this degree of specialization was practical in small Paleolithic groupings. As an example of the efficiency of large-scale specialization, consider the fact that it took about 10 square miles to feed a single ancient hunter-gatherer, but after the Neolithic (Agricultural) Revolution, it took only one square mile to feed 50 people.

As we shall see, all this efficiency comes with a price. This market economy, with its specialization and diversification, results in greater economic output, but also greater socio-economic inequality than we would find in a hunter-gatherer economy. The GDP of the market represents the involvement and effort of a great many people, and as such represents the bounty of all the market's participants—the group's. The problem with modern market economies is how to equitably divide up the spoils of such an efficient and productive system.

Rules, Regulations, and Cooperation

A modern market system, with all of its diversification and specialization, and with all of the buying and selling that goes on, is extremely complex, and one that requires thousands of rules, regulations, and laws. There are village laws, city laws, state laws, and federal laws—too many to count. This network of ordinances is the platform upon which the market operates, and without which the market would descend into chaos. Market interactions represent an exercise in cooperation, and could not proceed without an adequate degree of trust and acquiescence on the part of its participants. The level of cooperation demonstrated by the participants of the modern market is indicative of both their desire to succeed as individuals and their instinct to survive as a group. Everybody needs the system to work. So essential and pervasive are the workings of the modern market in our daily lives that it would be hard to imagine civilization without it.

Market Pricing

Let us look at some other characteristics of a market-type economy. Previously mentioned was the concept of intra-species competition and conflict. We know that when any species grows in number, and when its population density increases, competition for the limited resources within the pertinent eco-system will result. It is easy to imagine how this competition would play out in ancient times with competition for game and edible plants. However, it is a little less obvious how intra-species competition and conflict plays out in a large, complex, modern market economy—but it is just as real and just as consequential nonetheless.

Ideally, each of us has a skill or a product that we bring to the market, and in return we try to get as much product and service from the other market participants as we can. It is only natural for each of us to try to get the most in return for what we have to offer. In essence, we are trying to maximize our survival chances by getting the largest return from the market.

This competition—this push and pull—plays out every minute of every day in the marketplace. The market is very large and complex, and because of its size and complexity, it can obscure the interpersonal battle for survival.

At any given time, any society produces a certain totality of goods and services—its GDP. Just how much of that GDP "pie" is each individual entitled to? What fraction of the pie can an individual wrestle or demand from the market—from the rest of society?

For example, if a barber charges $15 to cut someone's hair, one might say that the barber can claim $15 worth of meat, cheese, or gasoline from the market at large. That is his little share of the GDP for that one haircut. But it's a competitive world out there. Human nature being what it is, that barber might try to get a little more meat, cheese, and gasoline for that haircut by charging $20 for that service.

Or imagine that biking has become very popular, and a bicycle manufacturer raises the price for a bike from $100 to $125. Each bike sold at the higher price represents a slightly greater share of the GDP that the bicycle manufacturer has wrangled from the rest of the participants in the market.

By contrast, if biking suddenly becomes very unpopular, the price of that bike may drop to $75. Consequently, the manufacturer will now share less of the GDP, and the other market participants will have gained share. We should mention at this point that competition between barbers and competition between bike manufacturers will mitigate extreme price gouging, but everyone will still try to maximize profits.

Intra-Species Conflict

This tussle happens over and over again. Teachers negotiate for higher wages. The grocer raises or lowers his prices. The cost of a barrel of oil changes day-by-day. The landscaper adjusts his rates. Sometimes workers even go on strike for higher pay. In each case, each market participant tries to extract from the other participants the maximum price or the maximum wage for his particular goods or service. This is intra-species competition and conflict played out in a market system. It is a seemingly bloodless intra-species struggle, but it reflects the individual's drive to survive, and at times it can get downright predatory.

A case in point is the pharmaceutical industry, which charges as much as it can for various drugs—many of them life-saving medications. The people who require these pharmaceuticals to maintain their health have very little choice but to pay the cost of the prescription. In this particular case, intra-species conflict can have life and death consequences. But, because modern societies are large and impersonal, people aren't necessarily aware and conscious of the adversarial nature of the market.

If we look around we can see other examples of intra-species conflict. People sometimes wrestle over a sale item on "Black Friday," or there may be a little pushing and shoving at the supermarket. Sometimes people resort to crime to survive—they rob stores and other people. They sell drugs and get involved with various other illegal activities. As a result of this competition and conflict, we see the necessity and importance of rules, regulations, and laws. We see the necessity of a police force and a judicial system, whose job it is to enforce those laws and resolve that intra-species conflict without bloodshed. In *paleo times,* if someone attempted to hoard all the supply of game, a vicious life-threatening fight would have probably broken out. Today, resolution of such conflict would hopefully be resolved peacefully, and if necessary, with the help of the police and the courts.

On a larger scale we can see the effects of intra-species conflict when we look at the causes of revolutions, civil wars, and wars between nations. Humans go to war because of disputes involving land, oil, water, and mineral rights. There is also so-called "class warfare," when the rich are at odds with the poor. Intra-species conflict and competition—this power struggle— plays out in the streets, the workplace, the marketplace, and in politics each and every day. This is especially true in tough economic times, when the strong take advantage of the weak.

> *"There's class warfare, all right, but it's my class, the rich class, that's making war, and we're winning." – Warren Buffett (American investor and philanthropist)*

The Invisible Hand

The term "the invisible hand" was introduced by Adam Smith (Scottish economist, philosopher, and author) in several of his writings during the eighteenth century. The term has perhaps been overused and generalized from Smith's original intention. Smith used it as a metaphor to explain how seemingly self-centered actions by individuals in the market can often result in the common good.

For example, an individual, in an effort to make a lot of money, might decide to start a business in town. This new business might provide a new product or a much-needed service to the area, and also employ several of the town's residents. This would be a "win–win" situation—the entrepreneur would benefit, but so would the community as a whole.

Nowadays, the term is sometimes also used to explain how prices and wages are arrived at for the various goods and services in a market. Some economists who advocate for unregulated and unfettered markets sometimes use the term as if free market economies are imbued with some sort of a philanthropic or divine force—an *invisible hand*—whose guidance results in the common good. Some may even use this concept as a rationale for greedy behavior, claiming that if everybody just did what was best for themselves, society as a whole would be better off.

We know of course that there are no external, beneficent forces watching over and guiding the market. Market dynamics are instead the result of each of us trying to satisfy our wants and needs. What drives the market is not some invisible force, but human nature—man's drive to survive. Prices and wages are determined by the push and pull of this rather impersonal intra-species competition, not some beneficent *invisible hand*.

Any of us who have taken Economics 101 is familiar with the "supply and demand" curves for any product or service, and how the intersection of those two curves determines price. *The Paleo Perspective* argues that the intersection of those supply and demand curves is a function of intra-species competition.

Capitalism

So far, we have been talking about market economies, regardless of whether those markets are owned and controlled privately or by a government agency.

To simplify things a bit, let us just say that in a purely capitalistic economy, everything is owned and controlled by private individuals, whereas in a purely communistic society, everything about the market is controlled and owned by the state. In reality, just about every economy in the world is a combination of both private and state control. Some countries, of course, are further to one end of the economic spectrum than others. Let us focus on markets that are mostly capitalistic and therefore privately owned and operated.

Capitalistic markets represent a major divergence from the Paleolithic economy of our hunter-gatherer ancestors. The primary focus during *paleo times* was on survival through group cooperation. One might imagine that life had a rather communal feel to it. People were very dependent on each other, and an individual had little chance of surviving without the help of fellow group members.

However, after the Neolithic Revolution, human society became fairly well established, and the survival of our species was virtually assured. Feeling a bit more secure, the individual could now focus on increasing his or her individual survival chances within the established social structure. It wasn't so much whether or not the human species would survive, but how an individual member of the species could survive in the rough and tumble of a market economy.

The capitalistic market economy emphasizes private ownership, not group ownership. It emphasizes individual success, not group success. (Socialism, in general, is focused more on group success.) The American economy and culture are good examples of the exultation of the individual over the group. One might suggest that American capitalism is a pivot away from the celebration of communal living, and a move towards the aggrandizement of the individual.

Market-Based Solutions

Capitalism's focus on the satiation of individuals' needs and desires puts it very much in sync with a notable aspect of human nature. This is particularly true in a country like the United States, where consumerism is paramount and accounts for approximately three quarters of the GDP. Understanding and acknowledging the relationship between capitalism and human nature should motivate policy makers to seek market-based solutions to social and economic problems whenever possible. While market-based solutions aren't possible or practical in every situation, there are many times when they can be employed quite effectively.

For instance, giving people (especially the rich) a tax deduction for contributing to charities has worked quite well. The rich see these tax deductions as a way to reduce their tax burden and maintain their wealth. Maintaining one's wealth satisfies the desire to insulate oneself from life's risks, thereby enhancing one's chances of survival. Let us not forget that the drive to survive is an omnipresent force that is deeply encoded in our DNA.

Another example of a market-based solution is giving people tax credits to purchase insulation or energy-saving windows for their home. This government policy encourages individuals to consume less fossil fuel, thereby helping to curb pollution and reduce global warming. Similarly, reducing electricity rates or mass transit fares during off-peak hours encourages more efficient use of these services. Offering a tax deduction for children, interest on a home mortgage, or retirement savings accounts are also examples of market-based solutions designed to steer individuals toward socially desired goals.

Generally speaking, a free market economy allows for individuals to problem solve and make decisions regarding their needs, and generally outperforms a state-run economy. Modern economies are too large and complex to be micro-managed by the state, especially if the state is run by a single dictator or a dictatorial coalition. If we think of this in terms of an ancient hunter-gatherer band, it would be like expecting the leader of the group to make every decision regarding the day-by-day, minute-by-minute activities of each and every group member. It is much better for individual group members to figure out for themselves what they need to do to survive, and what they need to do to be a productive member of the group.

Market Anonymity

As previously stated, most modern markets are large and complex. Because of its size and complexity, each participant enjoys a degree of anonymity in the market. Each participant enjoys the "benefit" of having a degree of separation between himself and the other market participants. This is in sharp contrast to the more intimate Paleolithic grouping.

The owners of a big company, or the shareholders of a corporation, are not intimate with, and do not necessarily come face-to-face with all of their customers. It's hard to feel guilty about getting the most money you can out of your fellow man when the transactions are so impersonal. The owners and shareholders are just trying to maximize profit and their market share to enhance their viability.

When a company pollutes a nearby river, the owners and the workers do not necessarily see firsthand how the people downstream are adversely affected. The company understands too well that it is cheaper to discharge waste without treating it first. Because of the separation between the factory and the environmental effects—a hundred miles away, and perhaps not to be felt for another twenty years—it is easier for the polluting business to ignore the negative consequences of its actions.

Does a deli owner feel guilty because he has raised the price of a ham sandwich? Probably not. Yes, he may know some of the customers personally, but there is also the less well-known public at large that he does business with. And the deli owner is probably thinking that he too is entitled to make a living, just like his customers. Everybody is trying to wrangle a larger share of GDP for himself or herself from everybody else. Everybody is just trying to make a living. That's how the business world works. That's how the market works.

Consider another case in point. If we look at the price of some drugs—drugs that treat serious diseases—we can see examples of drugs whose prices are unnecessarily high. Yervoy, Opdivo, and Keytruda are used to treat certain cancers. Their cost to the patient exceeds $120,000 per year. A drug named Cerezyme, used to treat Gaucher disease, and a drug named Kaldeco, used to treat cystic fibrosis, each cost over $300,000 per year. According to an article in *The New York Times* (September 9, 2015), written by oncologist Ezekiel J. Emanuel (of the University of Pennsylvania), these drugs are significantly overpriced. They are overpriced even when you consider the cost of research, testing, and development.

One reason certain drug prices can be so high is the lack of competition from other drugs that do the same thing. They simply might not exist at any given point in time. Another very important reason is that the patients who are stricken with these serious diseases will pay anything for relief. The makers of these drugs could charge less for their products, but given the lack of competition, and given how desperate these patients are, pharmaceutical companies often act to maximize their profits.

Indeed, the very concept of a "business" seems to absolve the owners of any moral responsibility they might have toward the rest of society. After all, the primary purpose of a business is to make money. Here too, there is a degree of anonymity and separation that makes it easier for pharmaceutical company officials and workers to not focus on the personal hardships that the high prices of these drugs cause.

In contrast, social interactions in *paleo times* were more personal, evoking more compassion, sympathy, and empathy. It is hard to imagine a Paleolithic group member withholding help for a sick cohort. In a large impersonal market system, these emotions are less likely to kick in. And so pharmaceutical companies tend not to feel guilty when they maximize profits at the customer's expense.

The Corporation

The existence of the corporation in our market system serves to further insulate individuals from the effects that their actions have on their fellow human beings.

Corporations have been around for centuries. They are legal entities, separate and distinct from their owners in many critical ways. However, corporations do enjoy most of the rights and responsibilities that a citizen possesses. For example, a corporation has the right to enter into contracts, loan and borrow money, sue and be sued, hire employees, own assets, and pay taxes.

The corporation has a life of its own, and can outlive its creators. It can take on the role of a social being—a person—entitled to some of the same civil rights as any citizen. In fact the Supreme Court in an 1886 decision (Santa Clara v. Southern Pacific Railroad) ruled that a private corporation was a "natural person." More recently, the Supreme Court (Citizens United v. Federal Election Commission, 2010) concluded that corporations, just like people, were entitled to free speech under the First Amendment to the Constitution.

A critical feature of a corporation is its limited liability. That is to say that corporate shareholders have the right to participate in the profits, through dividends and/or the appreciation of stock, but are not held personally liable for the company's debts. There are thousands of corporate rules, regulations, and laws that govern the behavior of corporations. Most of these rules were written by corporate lawyers and lobbyists, and so it should come as no surprise that the vast majority of these laws, regulations, and rules are of great benefit to the corporate world.

The Corporate Citizen

Society can come to think of a particular corporation as having a certain character or temperament. Indeed, a corporate business can possess a distinct culture. Google, Boeing, and Wal-Mart each have a distinct corporate culture and persona. Corporations can be good citizens, but they also can be mercenary and heartless. Let us not forget that the primary purpose of a corporation is to make money for its shareholders.

Quite often people associate various socially unacceptable behaviors with a corporation, but do not hold the stockholders and employees socially responsible for such behavior. And stockholders and employees can absolve themselves of guilt by reasoning that it is not them but the corporation that engages in misdeeds. If an individual owns a few shares of DuPont, he or she generally finds it easy to not take responsibility for any misdeeds that DuPont might have participated in. Tens of thousands of Exxon employees do not go to bed each night feeling guilty for an oil spill—such as the one in Prince William Sound, Alaska in 1989.

A business or a corporation can provide us with a degree of separation, lessening our chances of experiencing genetically appropriate shame and guilt. And so the existence of corporations makes it easier for us to avoid taking responsibility for the consequences of our intra-species competition with our fellow man.

We can perhaps think of the corporation as a "straw man"—a stand in for human cupidity and ambition. It provides a wall, a curtain, a degree of separation between the individual and his fellow market participants. Human wants and desires are behind that corporate wall—much like the man behind the curtain in the children's novel *The Wizard of Oz.*

Compare this lack of culpability to the guilt and shame a hunter-gatherer might have experienced if his actions harmed his fellow clansmen. In *paleo times,* humans lived in groups of perhaps 20 to 60 (maybe as many as 100) people. Most individuals were either related to or familiar with each other. In such small groupings, human emotions such as compassion, empathy, shame, and guilt can do what they were designed to do—to mitigate man's penchant for selfish behaviors.

The creation of the corporation—an inorganic entity void of human emotion—would have seemed unnatural in Paleolithic times. It would have been viewed as some sort of a gimmick—a contrivance to distract man from his natural sense of right and wrong.

Market Ethics and Morality

If we think back to the chapter on morality and ethics, we are reminded that the basis of human morality is the Golden Rule—concern for our fellow man. And, of course, concern for your fellow man (and for society in general) is a core principle of most modern day religions. But the essence of capitalism is the private production and ownership of material goods and services, not the welfare of fellow human beings. The focus of capitalism is more self-serving than altruistic.

When we use the expression "getting ahead in life," just whom or what are we getting ahead of? It would seem that we are really expressing a desire to improve our own lot without focusing much on how other people are doing. At times, the financial rewards of capitalism can tempt people to engage in socially reprehensible behavior. And one could argue that selfishness is becoming an ever-more accepted value in today's society.

Hollywood's commentary on modern society's acceptance of selfishness and greed was expressed in the 1987 film *Wall Street*. The expression "greed is good" was made infamous in a speech by lead character Gordon Gekko (played by actor Michael Douglas): "The point is, ladies and gentleman, that greed, for lack of a better word, is good. Greed is right, greed works." Most people who have seen the film come away with an understanding of just how corrosive a culture of greed can be on society.

Let us consider a few fairly recent examples of just how immoral and ethically challenged some individuals and corporations can be. The following is an excerpt from an article appearing in *The New York Times* on May 31, 2013. The article was written by Niki Kitsantonis.

> *ATHENS - Nearly two years after losing his job as a car salesman and with bills and debts piling up, Angelos started surfing the Internet for postings outside Greece. A small ad on a Web site offering "opportunities abroad" caught his eye, and he dialed the accompanying contact number and was told about a factory job in Sweden.*
>
> *A month later, he was out $2,300 and still jobless.*
>
> *"They told me to wire the money to cover procedural costs and the airfare," said Angelos, 38, a father of two who declined to give his full name for fear of jeopardizing future employment possibilities. The airline ticket never arrived in the mail, and follow-up calls went unanswered. A 300-mile road trip from Athens to the northern port of Thessaloniki, the job agency's stated location, led nowhere. The address did not exist.*

Another example of questionable business ethics can be illustrated with the bankruptcy of Enron Corporation in 2001:

> *Enron's downfall, and the imprisonment of several of its leadership group, was one of the most shocking and widely reported ethics violations of all time. It not only bankrupted*

the company, but also destroyed Arthur Andersen, one of the largest audit firms in the world. Enron C.E.Os, Kenneth Lay and Jeffrey Skilling, were tried together on 46 counts, including money laundering, bank fraud, insider trading and conspiracy. Skilling was convicted on 19 counts and sentenced to over 24 years in prison. (Forbes, February 2013)

A third example of unethical business behavior can be found in the rather famous case of Bernie Madoff:

In 2008, financier Bernard Madoff was arrested at his New York City apartment and charged with masterminding a long-running Ponzi scheme later estimated to involve around $65 billion, making it one of the biggest investment frauds in Wall Street history. Madoff launched an investment-advisory business that had become a Ponzi scheme, in which he paid his earlier investors with funds received from more recent investors. For years, clients of this business were sent account statements showing consistently high and fraudulent returns. Potential new customers clamored for Madoff to invest their money. In 2008, with the U.S. economy in crisis, Madoff's financial swindle began to fall apart as his clients took money out faster than he could bring in fresh cash. On March 12, 2009, Madoff pleaded guilty to the 11 felony counts against him, including securities fraud, money laundering and perjury. On June 29 of that year, a federal district court judge in Manhattan sentenced Madoff to 150 years behind bars, calling his actions "extraordinary evil." (History.com)

Here are three more quick cases in point reported to us by Paul Krugman in an op-ed piece in The New York Times (September 15, 2015):

Item: The C.E.O. of Volkswagen has resigned after revelations that his company committed fraud on an epic scale, installing software on its diesel cars that detected when their emissions were being tested, and produced deceptively low results.

Item: The former president of a peanut company has been sentenced to 28 years in prison for knowingly shipping tainted products that later killed nine people and sickened 700.

Item: Rights to a drug used to treat parasitic infections were acquired by Turing Pharmaceuticals, which specializes not in developing new drugs but in buying existing drugs and jacking up their prices. In this case, the price went from $13.50 a tablet to $750.

Finagling the books of a large corporation, cheating people out of money by operating a Ponzi scheme, circumventing auto emission standards, selling tainted food products, and price gouging on prescription drugs are all examples of immoral and unethical behavior—perpetrated by individuals, businesses, and corporations.

Not to excuse or condone these behaviors, but one could argue that these misdeeds were not perpetrated by inherently evil individuals, but rather by people whose conscience and moral compass proved inadequate in dealing with a large and impersonal marketplace. We should understand that the last 10,000 years or so have brought such drastic, momentous, and far-reaching changes to human society that in many ways we are in uncharted waters. Our brains were wired to promote ethical behavior in smaller, more personal groupings. Our conscience was designed for Paleolithic times, not modern times. Human nature hasn't changed, but human society has.

This conundrum reminds us of the fact that evolution has endowed humans with an instinct for both competitive and cooperative behavior. Unfortunately, the impersonal nature of modern society presents us with circumstances that seem to bring out the competitive instinct more than cooperative one.

Wealth and Income Inequality

As previously pointed out, the Neolithic Revolution precipitated the growth of villages, towns, cities, and great kingdoms. The increase in size and complexity of communities allowed for the accumulation of great wealth and power in the hands of the rulers of those social entities. Kings had it in their power to subjugate those living under their rule—to tax them and to coerce their labors. Kings, pharaohs, and emperors grew unimaginably rich, forcing their subjects to build for them great palaces, castles, monuments, and elaborate tombs. This was not so in Paleolithic times, when groupings were small and less complex. No leader of a hunter-gatherer troupe had the opportunity and power to subdue, coerce, and rule over tens of thousands of his fellow human beings.

Besides the Neolithic's spurring of municipal growth, it encouraged the establishment of a market system. To be fair, not only are market economies not inherently evil, but they can in fact offer many social benefits. Many times self-serving behavior on the part of the individual can improve the living conditions of other members of the group, and the group as a whole. (This unintended positive consequence is part of the concept of the aforementioned "invisible hand.")

Additionally, because of specialization, the large-scale production of goods, and the large-scale delivery of services, modern market economies are very efficient and can generate a lot of wealth. With each person contributing to production via his or her specialty, much can be produced. The problem becomes how to distribute the fruits of this specialization and mass production—which is in many ways a group effort— in a meaningful and equitable way. Intra-species competition and conflict will tempt each individual to exercise as much leverage as they can to maximize their share of the group bounty—to maximize their income and enhance their own survivability.

The capitalistic nature of modern markets can be particularly problematic. The people who own the means of production, such as the factory owner, the business owner, and the venture capitalist—all those in positions of power—will, of course, get the biggest shares of this mass-generated bounty. One might use the term "economic leverage" to help describe this phenomenon. Economic leverage can be compared to mechanical leverage (like a crowbar or a pulley) in that it multiplies one's economic power.

Let us imagine a single unskilled worker on the floor of a factory that happens to employ about 100 people. Any one of the individual unskilled workers has very little economic leverage because he or she could easily be replaced by another unskilled worker from outside the factory. He or she is not in a position to demand a lot more money. On the other hand, the CEO of this business has a lot more economic leverage. That CEO can demand a much higher salary because he or she is more highly skilled and cannot easily be replaced. The owner of the factory, with all its machines and equipment (capital), has the most economic leverage, because without the owner, the factory and the business would cease to exist. (One should note that if the workers formed a union they would drastically increase their bargaining power—their economic leverage.)

One could argue that without the workers, or the CEO, or the factory itself, nothing would get produced, and therefore nothing would get added to the GDP. The problem is how to equitably distribute the added share of the GDP produced by that factory given the fact that the worker, management, and owner each have a very different degree of economic leverage. We should note that although a good deal of cooperation is necessary for a market system to operate, intra-species competition is very much at the core of the market's dynamics.

Because of the nature of a capitalistic market, and because human nature is what it is, some rules and regulations to mitigate excessive human selfishness and greed are very much called for. We should recall the concern that Pope Francis has expressed regarding unbridled capitalism, the plight of the poor, and growing income and wealth inequality.

The market, if left on its own, can result in significant income and wealth inequalities. We have observed this in the past during the so-called "industrial revolution," when industrialists such as J.P. Morgan, Andrew Carnegie, and John D. Rockefeller amassed great fortunes, while the working class struggled to make ends meet. Many people see similarities between the past and the present.

"Some say we are presently experiencing another Gilded Age where for the first time since precise recordkeeping began a century ago, 10 percent of Americans take in more than half the country's income. In the last 40 years, the income of the top 1 percent of Americans has quadrupled, while incomes for everyone else have stagnated. Billionaires such as the Koch brothers and Michael Bloomberg, empowered by the Supreme Court, spend fortunes to influence politics, just as J.P. Morgan and California railroad magnates once did. Study after study shows that we are in the midst of a new Gilded Age, in which a yawning, gold-plated gap between the richest and the rest of us risks collapsing the American ideal of fair play and democracy itself." (Howard

Fineman, "A New Gilded Age Threatens The State Of Our Union," *Huffington Post,* January 23, 2014)

A market economy has a natural drift toward inequality. Anyone who has ever played the popular board game "Monopoly" comes to realize that given enough time, someone is going to own all the properties on the board. In the real market, similar dynamics are at work. Wealth tends to generate more wealth. Once you have made your first million, it is considerably easier to make your second million. Wealthy people can afford to give their children a good education, and have an important network of social connections for their family members to take advantage of. They live in good neighborhoods, attend good schools, and have good social connections. In contrast, poor people often live in crime-ridden neighborhoods, attend failing schools, and associate with all the "wrong" people.

> *"Wind is to a sailor is what money is to life on shore."*
> *- Sterling Hayden (American actor and author)*

Several economists and sociologists have recently made reference to the correlation between zip codes and socioeconomic status. In his book *Coming Apart* (2012), Charles Murray discusses elite neighborhoods and "super zip codes." The basic idea is that living in the right zip code correlates with better educational and career opportunities, safer neighborhoods, and enhanced wealth. Because of this correlation, one's likelihood for success can be likened to some sort of economic matrix, where your starting position on this matrix has a lot to do with your overall success in life. Perhaps being born in the right zip code can be likened to winning some sort of socioeconomic lottery.

If we look at some data (National Bureau of Economic Research, 2014) on wealth distribution in the United States, we discover that the top 10% own 75% of the wealth. The top 1% own 43% of the wealth. And the top 0.1% own 22% of the wealth in the USA. Looking at it another way, the top half of America owns about 98.9% of the wealth, while the bottom half owns only 1.1% of the wealth (Eric Zuese, *Washingtonblog.com*, July 7, 2015).

From these facts we can conclude that although a capitalistic market economy can generate great wealth for a country as a whole, that wealth tends to be very unevenly distributed. This is in sharp contrast to a hunter-gatherer economy, where there wasn't much wealth, but whatever wealth existed was more equitably distributed. Indeed, the very concept of accumulating wealth is relatively new. There was no such thing in the Paleolithic world.

According to the US Census Bureau, the average income per person in 2008 was approximately $27,000, and it hadn't changed that much by 2016. Consider if you will, the fact that major sport figures, entertainers, and corporate CEOs earn annual incomes in the millions. Are they really a million times smarter, stronger, or more talented than the average guy? Of course not. If these wealthy individuals had lived in Paleolithic times, they might (or not) have turned out to have been the head of their particular Paleolithic clan, but they would not have been a million times wealthier than any member of their respective clans. This discrepancy in income and wealth is a byproduct of the modern capitalistic market system.

Most would agree that while capitalism is capable of creating tremendous wealth, it is not good at distributing that wealth in a very equitable manner. The natural equilibrium state of an unfettered capitalistic market economy is not the equal distribution of wealth, but rather the concentration of wealth in the hands of a few. Money makes money. Left unchecked, plutocracies and monopolies will emerge.

The existence of paper money itself allows for the accumulation of wealth. A great deal of it can be stored in a relatively small space, and it is extremely portable. Furthermore, in the age of electronics, money is just a number on a page or on a computer screen. Even if he had the opportunity to do so, Paleolithic man could not carry with him any sizable quantity of material goods, and therefore could not accumulate significant material wealth. Nowadays, one need only carry one's checkbook or credit card. Furthermore, "money" itself can become a commodity—a piece of merchandise.

Consider the recent emergence of the "bitcoin" and other virtual currencies (referred to as altcoins). The value of a bitcoin went from $1,000 to $19,000 in 2017. Note also that the value of the US dollar also changes in the world marketplace—on a daily basis. And so "money" itself has moved beyond its initial raison d'etre—as a medium of exchange— and has become a commodity to buy and sell and to accumulate.

The Market, Technology, and Employment

In a way, the market is a victim of its own success. Diversification and specialization have sparked tremendous industrial and technological progress. Mechanical and electrical innovations have revolutionized how we produce the goods and deliver the services necessary for our survival. Technology can put some people out of work (replaced by a machine or robot), but at the same time, create opportunities for others (engineers and computer programmers).

Take, for example, the printing industry. Before the invention of the printing press, all copying was done by hand—mostly by clerics. Modern printing machines and techniques can produce thousands of copies in minutes, using less people, and with very little associated cost. And so there is no need for armies of people to copy documents, but there is a need for engineers to design copy machines, and for individuals with the technical skills to keep them running smoothly.

Or consider the software engineer who designs a program to help us file our income taxes. A program like this can put lots of accountants out of work, but it does encourage people to become computer software designers. In general, the development of the computer continues to revolutionize industry in every conceivable way.

Are all technological developments good for society? Clearly, some of them are. The development of various vaccines, surgical techniques, and safety devices are viewed by most as examples of true human progress. But other technological advances, like more lethal weaponry, processed foods, and pesticides can have negative consequences as well. The new mobile phones are amazing and helpful devices, but they also can be the cause of automobile accidents. There is also the question of whether such devices make us more social or less. Do they represent a net positive for society or a negative? Only time will tell.

Like it or not, technological advances continue day-by-day, with no end in sight. Arguably, some are good and some are not. What effect do they have on the labor force? What effect do they have on the unemployment rate?

According to *Fortune Magazine* "The U.S. has lost roughly 4 million jobs since 2000 due to automation. Robots are replacing workers at an accelerating pace, and the share of tasks that are performed by robots will rise from a global average of around 10%—across all manufacturing industries today—to around 25% by 2025." (Wolfgang Lehmacher, "Don't Blame China For Taking U.S. Jobs," *Fortune.com,* November 8, 2016)

To say the least, technological progress can be an unsettling force, and can cause major disruptions in society and in people's lives.

Food, Clothing, and Shelter

With specialization and technological innovation comes the fact that less and less people are required to produce the basics necessary for survival—food, clothing, and shelter. In the *paleo economy* everybody was involved with securing these basics—there was full employment. In the modern market economy things are very different.

The invention of the cotton gin, reaper, thresher, steam engine, gasoline engine, and tractor (to name just a few) drastically increased our ability to feed ourselves. So efficient are today's farming techniques that we actually produce more food with much less labor than we did in the past. For instance, in the mid-19th century it took 25 men a full day to harvest and thresh a ton of grain; today one person operating a combine harvester can do it in 6 minutes. In 1870, eighty percent (80%) of the U.S. population was engaged in agriculture, whereas just under 2% of the workforce was involved in agriculture in 2008. And according the U.S. Census Bureau (2010), only 17% of the GDP was dedicated to residential housing, and a mere 2.8% of American household income was spent on clothing. These three statistics tell us that the three basic needs—food, shelter and clothing—can be provided for with much less than 100% of the American workforce. (It also indicates that most individuals do not have to spend their entire income on the basics.) If food, clothing, and shelter were all that the American economy was preoccupied with, then the majority of Americans would be out of work. And so the question becomes what else are Americans (or the citizens of any other modern market economy) involved with? How else do people earn a living?

Creating Jobs

There are other important industries that are arguably necessary and ancillary to the food, shelter, and clothing industries. The automotive, communication, and education industries are indispensable and crucial to these three basic industries in modern times. But there are also many industries that are conceivably unnecessary for human survival. The tobacco, gambling, and alcohol industries come to mind. One might even argue that these three particular industries are disadvantageous to our survival and have moral dimensions to them as well. We sometimes refer to these industries as "sin industries" and the taxes collected from them as "sin taxes."

Each of these sin industries is huge and provides many jobs for society. Local communities tax these businesses and use the money to fund education, police departments, and fire protection services—each of which is a socially positive use of tax dollars. However, each of these activities can be pursued in excess, and can lead to addictions. The human and social costs of alcohol, nicotine, and gambling addictions are well known.

The entertainment, sports, and toy industries do not seem to be counterproductive, but they are also not absolutely essential for survival. They do, however, serve to satisfy certain human desires. There are many occupations that are not absolutely necessary for survival but employ many people nonetheless.

Many times we hear economists and politicians talking about the need to "create jobs." We don't talk so much about creating jobs in the food, clothing, or shelter industries because the demand for these items is rather constant since everybody needs these basics. But we do talk about creating jobs outside of these basic industries. For example, a politician might advocate for the building of a casino because it will create jobs. Or a mayor may push to build a sports stadium in town to create more jobs.

Societies don't absolutely need a casino or a sports stadium to survive. The fact that we often choose the word "create" in these instances implies that society needs to devise a way to give people the opportunity to produce something, or provide some service, so that the individual will be entitled to a share of that GDP pie.

Was there a need to create jobs in a Paleolithic hunter-gatherer economy? Probably not. Most everybody would have been involved in securing the basics, and therefore would be entitled to share in the equitable distribution of those basics. There wouldn't have been a sizable portion of any Paleolithic group that was engaged in non-productive or non-essential endeavors that expected a fair share of the necessities (food, clothing, shelter) even though they did not work to secure those necessities.

With a big stretch of our imaginations (and sense of humor), we could possibly visualize these "non-productive" Paleolithic individuals being engaged in entertaining the rest of the group with jokes, tricks, or supplying the group with some sort of tobacco to smoke. Of course this difficult-to-imagine scene would only be possible if the productive part of the troupe had a surplus of food, clothing and shelter, and was willing to share and exchange the excess of these necessities for the various non-essential commodities—jokes, tricks, and smoke. Admittedly, this is a very unlikely scenario in a Paleolithic economy.

At any rate, our modern market-based economy, with its specialization and technological innovation, is quite efficient—much more efficient than a hunter-gatherer economy. This efficiency allows a good percentage of the population to engage in the production of many non-essential goods and services, and still share in the bounty. With an adequate supply of the basics (food, clothing, shelter), the exchange of these basics for all the non-essential and quasi-essential goods and services can proceed at full speed in the marketplace.

Consider one last point with regard to the issue of employment— or unemployment. When comparing employment in a market economy versus employment in the *paleo economy*, we must also be mindful of the fact that in the modern market, individuals must actively seek out their place in that market. Modern man needs to seek out a career, and get the necessary training and education required for that career. Securing a place in the market might require taking out a loan for a college education or to start a business. The modern market requires the individual to be proactive, a self-starter, and perhaps a bit of an entrepreneur.

By contrast, our hunter-gatherer ancestors did not need to look very far, nor be very creative, to find work and to secure a place in their economy. It was obvious what needed to be done and by whom. Modern man takes the dynamics of the market for granted—it was not always so. Circumstances are indeed very different.

The Market: Irrational and Complex

The market displays a certain degree of irrationality. It is often difficult to predict market trends and to fully understand market dynamics. What products will be valued and at what price? How much should a person be paid for his or her service? Are today's CEOs really worth many times what they used to be worth? Aren't they basically doing the same job as they used to?

It makes sense that the market should display a certain degree of irrationality and unpredictability because its players—human beings—can be irrational and unpredictable at times. Human nature can be both puzzling and frustrating. For instance, people can harbor racial, religious, sexual, or gender biases, which in turn can affect hiring practices and salaries. Female doctors and female computer programmers earn about 75% of what their male counterparts earn (Jeanne Sahadi, *momey.cnn. com,* April 12, 2016).

Additionally, the market is extremely complex, with many moving parts and with many interacting variables. Accurate analysis is difficult—arguably even more difficult than predicting the weather. Because it is both complex and a reflection of human nature, the management and control of the market cannot be done effectively by a single person, committee, or political party. This is why free market capitalism (with the proper governmental controls) usually generates more wealth for its citizens than a one party communistic system does. Communist China has seen significant economic progress with the introduction and toleration of some meaningful market practices. Old Soviet Union-style communism was very idealistic, and in many ways ran contrary to basic human nature. As a result, the Soviet system experienced significant economic troubles—troubles that eventually contributed to its final collapse in 1991.

Another aspect of a market's complexity is the interconnectedness of its various sectors. The performance of one sector can have a ripple effect on other sectors. For instance, if the lumber industry is adversely affected by weather conditions, then the price of wood may rise and eventually affect housing prices. High housing prices may in turn result in the need for fewer carpenters, plumbers, and electricians. Subsequently, layoffs in the construction industries could result in less demand for automobiles, refrigerators, and furniture.

We can see how problems in one sector of the market can cause a downward spiral in the economy as a whole. A case in point was the crisis that occurred in the subprime mortgage industry a few years back. This calamity started a cascade of events that eventually spiraled into the Great Recession of 2007. Conversely, if one sector of the economy is in an expansive phase, this may have a positive effect on other sectors, resulting in an upward spiral of the economy as a whole.

During a recession, governments usually try to stimulate one part of the economy, hoping to initiate an upward spiral. For example, the government could lower interest rates in an effort to stimulate home buying. This in turn could stimulate consumer purchases in related areas such as furniture, gardening, and lighting fixtures—resulting in a much-desired positive economic spiral.

An important consequence of this interconnectedness and complexity is the adverse effect that market fluctuations can have on people's lives. Individuals today are subject to powerful economic forces—forces well beyond their control. Recessions, depressions, and inflation are all part of life in a market system. People can fall victim to a system that lacks compassion, and seems to have a life of its own. This is in sharp contrast to a Paleolithic hunter-gatherer economy, where economic uncertainty was more related to the weather and other environmental circumstances than to a myriad of complex and often irrational market forces. Indeed, one could argue that in some ways, Paleolithic man was more in control of his destiny than modern man.

Modern Man in a Paleo-Situation

Let us now turn to several hypothetical scenarios, the purpose of which is to help us focus on several particular social and economic issues. As stated earlier, *The Paleo Perspective* represents a reductionist view. With that in mind, let us attempt to strip away some things that we normally take for granted in everyday modern life by imagining ourselves in a life-and-death situation. Hopefully, by imagining ourselves in a life-and-death situation, various socioeconomic practices will be looked at in a different light. Let us create a scenario that allows for our genetic disposition to come to the fore—the way it would have for our Paleolithic ancestors.

To accomplish this, the reader will be asked to imagine himself first in a lifeboat and then on a deserted island with a cadre of fellow survivors. It is perhaps easier for the reader to picture himself in these situations than in a Paleolithic setting, which he most likely knows very little about. Most people have read books or have seen movies where these two life-and-death scenarios are integral to the storyline.

It should also be pointed out that the use of these hypotheticals is similar to the use of ethnographic studies involving modern-day hunter-gatherer tribes. The fundamental question being explored in both cases is how modern man behaves in more "primitive" (survival) situations.

The Lifeboat

Let us imagine ourselves to be the victims of some sort of disaster at sea. We find ourselves in a lifeboat with a small (20 - 60) group of people. While in the lifeboat certain decisions need to be made: Do we distribute whatever food and water we have on board equally amongst the passengers, or are the more wealthy people entitled to a greater share of supplies? Would we think it is acceptable if the bigger, stronger passengers take a greater share of the supplies simply because they are physically capable of doing so? If there is any rowing to be done, do we expect everyone to contribute to the effort (according to his or her ability), or are the poorer passengers expected to physically work harder than the rich?

Most people when asked to imagine this scenario would feel that everyone in the boat would need to cooperate in order for everyone to survive. This cooperation would most likely include an equitable distribution of food, water, and rowing responsibilities. Life on the little boat would more resemble a socialistic commune than a capitalistic free-market society. Even if a great amount of time passed while in that lifeboat, we would not expect significant differences in the distribution of goods and services to develop. And so when we think about it, the economic scenario in the lifeboat would be very different than what we find in modern market economies.

We are asked to imagine this lifeboat scenario because in many ways it is similar to what our ancestors experienced in Paleolithic times. Our human ancestors lived together in small bands and in dangerous environments where cooperation was essential. And as described earlier, there would have been very little wealth or income inequality, just like our passengers in the lifeboat.

It is important for us to remember that human nature (our DNA design) was engineered for these relatively small, socially intimate scenarios in which it seemed quite natural for us to be both compassionate and egalitarian towards the other members of our closely-knit group.

The Island

Let us continue with our "survivor" scenario and imagine that our lifeboat makes it to some small deserted island, somewhere out in a vast ocean. We don't know exactly where we are, and we have little hope of being found and rescued. And just like life in the lifeboat, there are decisions to be made on the island.

Perhaps the most immediate and critical decision for the group to make is whether or not people should cooperate with each other—or is it every man for himself? In this life-and-death situation, most people would feel that everybody's chances of survival are greatly enhanced if everyone cooperates. Indeed, in this scenario, our genetic instinct to cooperate would be very strong. That is not to say that every single person would be cooperative. There might be a few individual outliers, but the cooperators would carry the day.

With 20 to 60 people on the island, it would quickly become necessary for the survivors to think about how they should go about organizing themselves. Should there be some type of democracy, with each person having a say in what happens? Or should there be some sort of dictatorship, formed around an "alpha" individual within the group? At any rate, there would need to be some form of governance, lest the island community fall into total chaos.

Whatever form of organization that developed, the decision about the distribution of food, water, and other vital supplies would present itself almost immediately upon arrival to the island. Is everyone equally deserving of supplies? Is everyone equally responsible to expend effort to do whatever is necessary to enhance survival chances—to look for food, water, and to build some sort of shelter? Would more talented people be entitled to a larger supply of food, water, or some other necessity? Or would there be an equal distribution of these necessities regardless of individual talents? Once again, if asked to imagine themselves in this life and death situation, most people would opt for the highly cooperative, egalitarian island society—at least at the beginning.

Patents

Let us now suppose that some time has passed and someone invents some kind of tool or device that enables him to collect more food—perhaps a tool for cutting or digging. Or perhaps some kind of weapon to make hunting easier, or a tool to help with the fabrication of shelters has been conceived. The so-called invention need not be a tool at all. It could be the development of a clever technique to get the coconuts out of the trees more easily, or a previously unthought-of farming technique. (This would be akin to the concept of intellectual property.)

Would the inventor feel that he was now entitled to a larger share of the supplies? After all, the invention did increase the food supply, or allowed for the fabrication of superior housing. Would the inventor think that he was entitled to some sort of "patent" on his product so that nobody else on the island could duplicate the invention without his permission or use it without paying him some extra fee for its use?

On our imaginary island, with its small group of closely-knit inhabitants, this kind of self-serving behavior would be seriously questioned, and would probably not be tolerated. In fact, this sort of self-serving behavior, if carried to the extreme, would probably feel like extortion to the rest of the island's inhabitants. In this situation, everybody understands that cooperation, not competition, would enhance the group's chances of surviving. This would be no time for intra-species competition and conflict.

But this sort of thing happens all the time in our present-day market system. We humans get new ideas all the time. New devices are invented and new techniques are developed. And when they are invented or developed, people and corporations obtain patents on their ideas and products, preventing other people from duplicating or using their product without permission—or without at least paying for the privilege of their use. With our modern system of patents, the inventors and designers have the potential to make a great deal of money.

There is no question that patents benefit the creator. But there is disagreement over whether patents, and the concept of intellectual property, benefit society as a whole.

The Paleo Perspective asks us imagine what it would have been like if our human ancestors didn't share ideas. What if one of our ancestors kept secret a new and more efficient technique for knapping stones (to create stone tools and weapons)? Or consider the necessary knowledge that needed to be shared and passed on with respect to the controlled use of fire. Think about the skill needed to skin an animal and make clothes, or the skill needed to build a hut, or to hunt effectively. What about the invention of the wheel, the bow and arrow, or the slingshot? It seems unlikely that the concept of "intellectual property" existed during Paleolithic times. It seems ridiculous to think that ancient man would have been better off if all of these important skills, ideas, and inventions were not shared freely, but instead were patented and hoarded for individual gain. And yet that is the way things are in modern times.

An interesting article concerning patent rights appeared in *The New York Times* (April 18, 2014). The article, written by Joe Nocera, refers to a book entitled *Birdmen* by Lawrence Goldstone. *Birdmen* is primarily about the origins of the airplane, but it also talks about an ongoing feud between the Wright brothers and Glenn Curtiss over competing airplane designs. The Wright brothers filed several patents regarding certain wing designs, and claimed that Curtiss was violating their patent. Years of lawsuits followed. According to Nocera: "There is no question what the Wright brothers sought: nothing less than a monopoly on the airplane business—every airplane ever manufactured, they believed, owed them a royalty. The Wright brothers were so caught up in litigation that they stopped innovating. By attempting to neuter Curtiss, the Wrights stifled the development of American aviation."

Along the same lines, consider the tremendous technological progress that the modern computer represents. One could not imagine all that progress being made if someone, or some company, had been allowed to patent the very idea of the computer. Indeed, one could argue that the free and open sharing of ideas is in fact much more conducive to the creative process than a system where everybody seeks a patent for their ideas.

However, there are those who would argue that patents are essential, that they enhance profits, and are therefore an incentive to innovate. One could feel this is perhaps particularly true for corporations, whose prime purpose is to make money for its owners and investors. In an effort to turn a profit, corporations make plans for product development, and can invest a great deal of time, manpower, and money in an effort to design and patent some new product.

While profits do provide reward for innovation, *The Paleo Perspective* would argue that innovation is ingrained in human DNA. Humans are creative by nature, and would invent and create with or without financial incentive. Indeed, one could argue that the individual scientists, chemists, and engineers that work for a large corporation have the urge to be creative regardless of who they work for. Creative people certainly want to earn a reasonable living, but if you gave them a 25% raise in salary, it is unlikely that they would be 25% more creative. One could also argue that people such as Bill Gates (Microsoft), Mark Zuckerberg (Facebook), and Steve Jobs (Apple) were creative by nature, and would have pursued their creative instincts regardless of any guarantee of the huge corporate profits that eventually followed.

There is also the idea that important innovations and inventions would have happened anyway, by a different person, at some later point in time. Do we really think that the wheel wouldn't have been thought of eventually? Computer software would have been developed without Bill Gates. The light bulb would have come along at some point in time without Thomas Edison. These were all inventions whose time had come—with or without the economic benefit of a patent.

Money, of course, does provide some incentive, but it is not the prime mover of creative thought. Creative individuals (as opposed to corporations) don't sit around thinking "what can I invent to get rich?" The fact is that creative thoughts are going around in creative minds all the time. Being greedy is not a prerequisite for being creative. Our genetically-driven creativity is not just there for the benefit of the individual. It is there for the group's benefit—for the species benefit—as well. Perhaps we can think of making a lot of money by being creative as the icing on the cake—but not the cake itself.

Private Ownership and Economic Rent

Let us now think about the idea of private ownership and something called "economic rent." What do economists mean when they use the term *economic rent*? By definition, it is the positive difference between the actual payment made for a factor of production (such as land, labor or capital) to its owner and the payment level expected by the owner, due to its exclusivity or scarcity.

For example, the price of a hotel room in downtown Boston, Seattle, or New York can be quite high—several times what the rate would be for the same room in a less exclusive area. Another example might be the high price of a life-saving drug. The price may be extremely high because the drug might only be available from one pharmaceutical company, or because it may be the only drug that has been shown to be effective against a particular life-threatening illness. This drug could possibly be very expensive to manufacture—but it also might be quite inexpensive to manufacture. A high price could be the result of a lack of market-based competition, and the fact that most people will do almost anything—pay any price—when they are confronted with a life-threatening medical situation. Similarly, if there is only one internet/cable provider in a particular area, there is very little incentive for that company to keep the prices for their service down. Indeed, the likelihood of some degree of *economic rent* occurring is extremely high due to the fact that a monopoly exists. Let's get back to our island.

As Time Goes By

Imagine some time has passed for our survivors on that island. Some shelters might have been constructed, some regular hunting might be taking place, and our inhabitants might be helping themselves to any fruits, berries, and edible plants that they could find on the island. If our survivors had come from countries with a capitalistic tradition, would they assume that the island community should also be run as a capitalistic, market-based community, with private ownership of resources? Consider the following questions, bearing in mind that upon arrival to that island, nobody owned anything on the island.

Who should own the most fertile pieces of land with all the fruit trees on it? Who should own the land with the only fresh water stream passing through it? Who should own the areas with abundant game for hunting? If the capitalistic system of private ownership were practiced on the island, the owners of these prime pieces of land could charge exorbitant fees (rents) for the valuable commodities present on their privately owned land. Indeed, the less fortunate inhabitants of the island might even wind up working for the fortunate few.

If we imagine ourselves to be on that island in this precarious situation, we might question the morality, or the righteousness, of private ownership of the island's natural resources. After all, our little group of survivors came upon this island—with its life-saving resources—by accident. And nobody had any claim to any land or resources at the very beginning of this adventure. It could be argued that any claims to private ownership of any of the various resources on the island would be viewed as selfish and even life-threatening behavior. Such self-serving behavior would have been a tough sell with our small band of survivors on that island, just as it would have been for a small band of hunter-gatherers back in Paleolithic times.

Gatekeepers and Middlemen

The term "gatekeeper" usually refers to a person in today's healthcare industry who controls access to various health care services. It is usually a doctor or an employee of a medical insurance company. Let us generalize the concept of a gatekeeper and apply it to our island scenario.

If your piece of property was in a particular location, whereby others on the island needed to cross your property to have access to a stream or access to the sea, you could act as a type of gatekeeper. Or perhaps people needed to cross your property to have access to good hunting areas. You could then charge people a fee to cross your land in order for them to have access to the water or hunting.

Similar to the concept of a gatekeeper is that of a "middleman." Imagine if you will, that kindling wood was in limited supply on the island, and everybody needed it to help start campfires. You could make it your business to collect and hoard this kindling material. Having done so, you could extract a high price for this limited resource. A ticket scalper is a modern example of a middleman. Nowadays, with the widespread use of the Internet, an individual or a group can buy up a large number of tickets —creating a scarcity—for an upcoming event before they go on sale to the general public. These tickets are then resold to the public at large for an inflated price.

In each of the aforementioned island scenarios, a person could take advantage of the economic leverage that he or she has over the other inhabitants, and benefit greatly. It should be noted that gatekeepers and middlemen do not add any economic benefit to the island community. They have simply taken advantage of various situations and used them to their own benefit.

The economic rent, gatekeeper, and middleman situations are similar in that they would allow a select few individuals to benefit at the expense of the rest of the island's inhabitants. In any case, given enough time, the result would most likely be a very lopsided distribution of wealth on the island.

It is difficult for most of us to imagine that on such a small island, with such a small, intimate band of people, its inhabitants would tolerate such self-serving behavior and the large degree of inequality that would result from such behavior. It is more likely that the island's resources would be shared more or less equally. A key point here is that this imaginary island scenario is much closer to the tribal living modality that probably existed in our evolutionary past than the large, impersonal, and complex market system we find ourselves in today. Primitive man's small troupes were similar to those survivors in the lifeboat and on the island in that they were in a dangerous and uncertain situation, where cooperation and harmony were essential. We must remind ourselves once again that our DNA was designed for very different and simpler times.

In Summary

The point of *The Paleo Perspective* is not to pass judgment on the pros and cons of capitalism, socialism, or communism. It does, however, ask us to understand and appreciate the fact that the economic realities that we find ourselves in today are relatively new, and that in many ways our genetic profile is mismatched with today's circumstances.

Ever since the Neolithic Revolution, with its technological innovations and market-based economies, society has been drastically altered. There were no such things as corporations or patents in Paleolithic times. Things like a first-class education, family wealth, high-powered business and social connections, and access to sophisticated technological advancements enable the "haves" to accumulate wealth orders of magnitude greater than the "have-nots." These success multipliers did not exist in man's Paleolithic past.

Was man designed to be a capitalist or a socialist? If we recall the fact that evolutionary pressure has exerted itself on humans on two levels—survival of the individual and survival of the group—we might conclude that man is a little bit of both. But we are a species designed in Paleolithic times, when man's instincts for selfish behavior were mitigated by the necessity for group cooperation. Our conscience—with its sense of fairness and its sense of right and wrong—and our capacity to feel shame, to sympathize, and to empathize with our fellow human beings were all designed to work best in those ancient small group settings.

Seeking individual wealth can be very enticing—even seductive. The modern market is better at facilitating the gratification of the wants and desires of the individual than it is at ensuring a just and egalitarian society. The application of a *paleo perspective* would suggest to us that some constraints (regulations) are necessary to mitigate some of the excesses that result from the mismatch between our Paleolithic design and modern circumstances.

> *"History has shown that where ethics and economics come in conflict, victory is always with economics."- B.R. Ambedkar (Indian jurist, economist, politician, and social reformer)*

CHAPTER 17

PALEO POLITICS

"Just because you do not take an interest in politics doesn't
mean politics won't take an interest in you."
- Pericles (Greek statesman and orator – 450 B.C.E)

This chapter on politics and the previous ones on religion and economics are particularly germane to the thesis and focus of this book. They are because only by looking through a Paleolithic lens can we better understand the nature of, the cause of, and the extent of the dystopian elements present in modern society. So let us now look at the topic of politics with a *paleo perspective* in mind.

Politics is so pervasive and omnipresent that it can fade from our consciousness. We sometimes forget how important it is and how it affects our daily lives in so many ways. Indeed, politics concerns itself with foreign policy, domestic policy, and the free exercise of our civil rights. Our American civil rights happen to include free speech, the right to bear arms, the right to freely practice religion, and the right to not be discriminated against because of gender, sexual orientation, disability, or national origin. Government policies affect banking regulations, taxes, healthcare, and police and fire protection. Food prices, clothing prices, home prices, and gasoline prices are all affected by local, state, and federal politics. Political machinations even affect the water we drink and the air we breathe.

In point of fact, there isn't much that isn't affected in some way by politics at some level. Politics is real, it is important, and it affects not only the quality of our lives, but it can also have a direct bearing on the difference between life and death itself.

> *"Politics is war without bloodshed, while war is politics with bloodshed."*
> *- Mao Zedong (Chinese communist revolutionary and*
> *founding father of the People's Republic of China)*

Politics, ostensibly, takes place on the surface, but underneath it is all about intra-species competition and conflict—survival. If we were still in a hunter-gatherer mode, politics might take the form of an argument, a little pushing and shoving, or perhaps even a deadly fight.

Today that struggle takes place not only on a person-to-person level, but also between groups. There are turf wars between gangs and battles for power between armed militias and the state. This battle takes place on a town, state and federal level, as well as between nations and between civilizations. Whether it's a clash between individuals, terrorism, civil war, or world war, it's all intra-species conflict. Man fighting man, groups fighting groups. We fight to survive as individuals and we fight to survive as a group. Politics is that battle, sometimes with bloodshed and sometimes without. Let us look at several specific examples of politics on the surface, but survival underneath.

Pocketbook Politics

Within the United States there is a battle being waged over gun ownership—specifically whether there should be restrictions, if any, on the type of weapon and ammunition that an ordinary citizen may possess. A case in point would be the ownership of military-type assault weapons such as the AR-15. The National Rifle Association (NRA) argues that the United States Constitution (in the Second Amendment) guarantees all individuals the right to bear arms, and there should be virtually no consequential restrictions or limitations on that right. One could assume that the NRA is simply defending a constitutionally guaranteed right. And while that may be true in part, there are other forces at work.

Many individuals contribute to the NRA, some of who are ordinary individual gun enthusiasts. But gun dealers, gun shop owners, and gun manufacturers also support the NRA. There is in fact an extensive network of businesses that depend on the manufacture and sale of firearms to survive. Besides the obvious examples of gun manufacturers—such as Colt, Berretta, Remington, and Smith and Wesson—there are the companies that manufacture ammunition, and the owners of firing ranges. There are the related industries of taxidermy, gun safety instruction, and hunting guides. There is the manufacture of sports clothing, targets, optical scoping devices, bulletproof glass, bulletproof vests, and gun safety boxes and cabinets. The list goes on and on. The point is that all of these people, all of these businesses, all of these manufacturers, and all of these industries have a stake in any laws or regulations that affect the manufacture, sale, and use of firearms. There are millions of dollars at stake. Millions of people make a living due to the existence and proliferation of firearms. This is how they survive as individuals, as businesses, and as a group. It is therefore no surprise that they are big supporters of the NRA. Their livelihoods depend on a robust firearms industry and their politics will follow their pocketbooks.

"It is difficult to get a man to understand when his salary depends on his not understanding it." - Upton Sinclair (American author)

As a second example let us look at the healthcare industry. America's national healthcare expenditure (NHE) was about 3 trillion dollars in 2012 (Dan Monro, "U.S. Healthcare Hits 3 Trillion," *Forbes.com*, January 19, 2012). Think about all the people, professions, and businesses that operate within the healthcare domain. There are doctors, nurses, and medical lab technicians. There are all those "non-professional" level individuals who work in hospitals, clinics, labs, and doctors' offices such as medical assistants, receptionists, and office cleaning personnel. Consider also the many companies that manufacture all sorts of medical devices such as pacemakers, prosthetic limbs, crutches, bedpans, etc. Let us not forget about the huge pharmaceutical industry, with over $900 billion in sales in 2015, and the medical insurance industry, which took in over $800 billion in 2015.

All of these people and businesses are affected by the many laws and regulations governing healthcare. Everybody from the top-notch surgeon and high-priced CEO to the blue-collar maintenance worker has a stake in the politics that governs the enormous healthcare industry. Even the people who own stocks in the healthcare sector of the economy are concerned about the surrounding politics. And just as with the firearms industry, the individuals and groups that make a living in the healthcare industry will form opinions, vote, and fight to maximize their salaries and profits. They will do this in response to their genetically-driven Paleolithic directive to survive—to survive as an individual and to survive as a group.

"Man is by nature a political animal"
- Aristotle (Greek philosopher and scientist – 350 B.C.E.)

The firearms and healthcare industries are no different than the agriculture, airline, or auto industries—or any other industry for that matter. People put food on the table, pay the rent, and send their kids to college by earning a living in one of America's many industries. They understand and are aware of the regulations, policies, and politics that affect their livelihood, and will act accordingly out of self-interest. The politics that plays out on the local, state, and federal level is just a manifestation of human intra-species competition and conflict.

Politics as a Profession

Politics is an industry, just like the beverage, petroleum, or clothing industry. Some people spend the majority of their adult life as an elected official or in some other capacity within the political system. There are mayors, governors, councilmen, assemblymen, congressmen, and senators. And there are lots of people that work for all of those officials.

As of 2015, there were 79 members of the United States congress that had served for over 20 years and 16 members that had served for over 30 years. The average age for a congressman is about 60, and most started their political careers as young adults. According to the Center for Responsible Politics (2012), the median net wealth for a U.S. congressman is a little under one million dollars, and the median net wealth for senators is around 2.7 million dollars. Politics is in fact a fairly lucrative profession. Besides the money, power and influence also come with the job.

Let us also not overlook the fact that many politicians go on to become well-paid lobbyists once they retire from office. Human resource departments from around the United States report that the median salary for a lobbyist is $108,727 dollars (*Salary.com,* January 2018).

> *"It has been said that politics is the second oldest profession. I have learned that it bears a striking resemblance to the first."*
> *- Ronald Reagan (40ᵗʰ President of United States)*

Politics as a Zero Sum Game

Unfortunately, elections are "zero sum" situations. There is only one president, one governor for each state, and only one mayor for each city. Similarly, the number of federal senators and congressmen is also fixed (100 and 435 respectively). When challenging an incumbent, one has to give the voters a reason to "switch teams." After all, there can only be one winner. The challenger is therefore tempted to vilify his opponent, or at least to minimize or deny any supposed success that the incumbent claims to have had. It is no surprise that most of us have heard politicians say and do whatever they need to in order to get elected. The result can be extremely nasty campaigns with lots of misleading and inaccurate information being tossed about. Lots of "fake facts."

And like the people in any other industry, the people in politics go to great lengths to keep their job, to protect what they have, and to pursue the rewards of their chosen livelihood. Their instinct for survival vis-à-vis their profession is just as strong as that for individuals in any other profession. Let us look at two examples of this survival instinct in action.

It is no secret that elected officials often depend on many contributions to seek office and to run an election campaign. In an article appearing in *The Huffington Post* (January 8, 2013), written by Ryan Grim and Sabrina Siddiqul, newly elected Democrats were advised (in a PowerPoint presentation by the Democratic Congressional Campaign Committee) to spend roughly 5 hours of every day soliciting for campaign contributions and doing anything else they could to help keep their elected office. That is roughly half of their workday. One can only assume it is no different for Republicans.

> *"Politics has become so expensive that it takes a*
> *lot of money even to be defeated."*
> *- Will Rogers (Actor, humorist, writer, and social commentator)*

Another case in point is the process known as "gerrymandering." Gerrymandering is the process by which election district boundaries are drawn in such a way to give one political party a majority in many districts, while concentrating the voting strength of the other party into as few districts as possible. Election district boundaries wind up with highly contorted and irregular shapes (resembling a salamander—hence the name). Both Republicans and Democrats are guilty of gerrymandering. This practice all but guarantees that incumbents will be re-elected. According to *Politifact*, 96.4% of incumbents were re-elected in 2014 in spite of the fact that for that same year the United States Congress had only an 11% approval rating. Similarly, in the 2016 elections, 97% of incumbent House members were re-elected, as were 93% of senators.

One could be cynical and think that any public good that a politician does is an unintended consequence of holding office, and that staying in office is really the primary goal. It reminds us once again that most people view their jobs as a means to make a living—a way to survive.

The Need for Government

> *"The most terrifying words in the English language are: I am from*
> *the government and I am here to help." - Ronald Reagan*

> *"The new rage is to say that the government is the cause of all our*
> *problems, and if only we had no government, we'd have no problems.*
> *I can tell you, that contradicts evidence, history, and common sense."*
> *- William J. Clinton (42nd President of the United States)*

The question of whether or not there is a need for government has been around for thousands of years. Here in America we seem to be consumed with the question of what the proper size and role of government should be. To put it mildly, Democrats and Republicans seem to have a difference of opinion on this question. What would a *paleo perspective* be on this matter?

It is widely acknowledged that man evolved as a hunter-gatherer, moving around in small, closely related bands of 20 - 60 people. Man instinctively understood that to survive, a certain level of group cooperation and organization was necessary. Most likely, some form of hierarchy was established—perhaps based on age, assertiveness, or leadership characteristics. Most would assume that for these small groups, this level of organization would have happened quite naturally, without the need for any formal voting. We might consider this hunter-gatherer group structure and hierarchy to be a very informal type of "governmental" organization.

You might recall that with the occurrence and spread of the Neolithic Revolution, the size and complexity of human groupings greatly increased. Towns, cities, and city-states require a more entailed and formal form of cooperation from their inhabitants than is necessary for a small Paleolithic troupe. Without a more structured form of cooperation, a state of chaos would descend upon any town, city, or state. This necessary, more formalized group organization represents a form of governance. It represents the establishment of government.

In modern times the concept of government includes many things. It includes a military, police and fire protection, an education system, health services, sanitation services, mail services, environmental protection services, public parks, bridges and roads, food and drug regulation, and water services—to name just a few.

Governmental laws, regulations, and taxes can sometimes be burdensome, but they are indeed necessary. In a way, government in most Western societies is so successful that it can become invisible and be taken for granted. Like oxygen in the room, it is not noticed until it is gone. People forget that if you take away one or two essential government services, turmoil is most likely waiting in the wings.

Consider a heavy storm or a blackout. Without electricity and gasoline, without access to roads, bridges and tunnels there is much panic to be had. People start to worry about food, water, safety, and crime. Things can become ugly very quickly, and everybody goes into survival mode. Fear of this kind of social breakdown is one reason why many people want to possess a firearm, or live in some remote area "off the grid."

Unfortunately, simply using the term "government" when discussing public services, or the people who perform these vital public services, tends to depersonalize their role in society. We don't tend to think of our friends, relatives, and neighbors as the "government." And so it becomes easy—even fashionable—to be suspicious and critical of the "government." Cheating on one's income taxes is thought of as cheating on the "government," not one's neighbors.

There is no question about the need for some level of government. And it is proper and necessary to criticize any government when there is inefficiency and corruption within that government. But it is counterproductive, and even destructive, to argue against the very concept of government. Government is necessary. The only question is how big that government needs to be.

In the United States the size and role of government is a hotly debated topic, especially between so-called conservatives and liberals. Generally speaking, conservatives favor a less involved, limited government, and liberals favor a proactive, more involved government. Is there an ideal size for government? In grappling with this question let us consider the extremes of each point of view.

Too Much Government

"A government big enough to give you everything you want is a government big enough to take from you everything you have."
- Gerald R. Ford (38th President of United States)

Let us start with the idea of a society that is virtually totally under governmental or state control. Let's imagine that a single dictator—perhaps a military dictator—rules a particular country. What would be the advantages of such a system?

One might assume that there would be much order in such a place. Perhaps a military-style police force with harsh penalties for infractions of the law would keep crime very low. Services such as transportation, health, sanitation, and power generation would all be run by the state. Additionally, the state would try to manage and control every aspect of the economy—making decisions about what goods and services should be produced. What are some of the drawbacks of such a totalitarian system?

One obvious drawback would be the lack of freedom enjoyed by its citizenry. And because of this lack of autonomy, creativity and innovative thought would most likely be curtailed. We would not expect such a country to be on the forefront of artistic, technological, or scientific achievements. In fact, autocratic societies tend to underperform in a whole host of social and economic areas such as per capita economic development, infant mortality rates, and life expectancy. It is hard to compare these developmental attributes universally across the globe because countries vary in size, climate, and location. And most importantly, different countries have access to very different natural resources.

That being said, it may be instructive if we compare North Korea to South Korea. Each country is roughly the same size, with the same climate, and each has access to roughly the same natural resources. What do we find when we compare several socioeconomic variables?

South Korea's infant mortality rate is 3.9 per thousand as compared to 24.5 for North Korea. South Korea's maternal mortality rate is 16 per thousand as compared to 81 for North Korea. Life expectancy for South Koreans is 80 years compared to 70 for the North. Virtually 100% of the South Korean population has access to sanitation facilities, whereas only 80% do in North Korea. And finally, the per capita GNP (Gross National Product) is $31,000 for South Korea, as compared to $1800 for North Korea (*The World Factbook*, published by the Central Intelligence Agency, 2011). There are more points of comparison, but most would agree that the people of North Korea, with their totalitarian, autocratic style of government, are not doing as well as the people of the democratic South.

In general, the totalitarian dictatorships around the world are less successful across a broad spectrum of measurements when compared to Western-style democracies. Consider the per capita GNP of the following dictatorships:

- Vietnam - $1,890
- Bangladesh - $1,080
- Chad - $980
- Haiti - $820
- Nepal - $720
- Uganda - $670
- Niger - $410

Consider the per capita GNP for the following democracies:

- Australia – $64,540
- Sweden - $61,600
- United States - $55,200
- Germany - $47,902
- Great Britain - $43,430
- France - $42,955
- Italy - $34,270

We should note that these democracies have market-based economies, where the majority of economic and day-to-day business decisions are made by millions of ordinary citizens—not by a single dictator. (The above figures were obtained from the World Bank, 2011 – 2015).

Modern economies are too large and too complex to be managed effectively by one person, two people, or even a committee of individuals. There are of course exceptions to every rule. When we look at the per capita GNP of Kuwait ($49,300) and the United Arab Emirates ($44,600)—both autocracies—we find that they are doing quite well. There are other examples of autocracies doing well, but we must consider the fact that many of these countries happen to be extremely rich in a particular natural resource—oil.

As stated earlier, it is hard to compare various societies around the world because of large differences in crucial circumstances. But there is a strong case to be made that too much governmental control can be detrimental to a nation's overall success.

Too Little Government

"We don't need a weakened government but a strong
government that would take responsibility for the rights of
the individual and care for the society as a whole."
- Vladimir Putin (President and Prime Minister of the Russian Federation)

In the United States we often hear some individuals on the "political right" advocating for smaller and smaller government. So let us now consider what happens when there is very little or no governmental control. Would this be a good thing?

Each year the United States think tank, "Fund for Peace," compiles a list that is called the "Fragile States Index." A "fragile state" has several attributes. Common indictors include a state whose central government is so weak or ineffective that it has little practical control over much of its territory. There is generally non-provision of public services, widespread corruption and criminality, refugees and involuntary movement of populations, and sharp economic decline. Generally speaking, societies that don't have functioning rules, laws, or regulations are failed states. Many of these countries are in a constant state of civil war, where local strongmen form militias and vie for control. Some of the nations at the "top" (doing poorly) of the 2016 list include South Sudan, Somalia, Central African Republic, Sudan, and the Democratic Republic of the Congo. It would seem that having too little government, just like having too much government, can be problematic.

Goldilocks-Size Government

In the children's story "The Three Bears," Goldilocks was looking for things that weren't too big and weren't too small, but were just right. Perhaps the same concept applies to government as well—too small and limited is not good, and too big and controlling is also not good. It would seem that we are looking for some kind of a "sweet spot" when it comes to the size of government. So let us look at some evidence for success in the middle.

In 2015, the United Nations published its annual *Human Development Report*. The UN looked at nearly 200 countries across a number of categories, including life expectancy, education, and financial wealth. The 10 countries with the highest ratings are Norway, Australia, Switzerland, Denmark, the Netherlands, Germany, Ireland, USA, Canada, and New Zealand. All of these countries are Western-style democracies with varying degrees of governmental control and varying degrees of socialism.

Let us take note that none of these 10 countries are considered to be autocracies or dictatorships. They are examples of countries with strong but limited central governments. Each of these nations exhibits a certain degree of socialism, with a substantial social safety net for its citizens. They represent governments that are not oppressive, yet strong enough to maintain order. And so success seems to be the ability to find the right balance between the two extremes—that sweet spot.

A Paleo Perspective on Government

The central question here is what type of organization (government) is most compatible with our DNA design? What would employing a *paleo perspective* suggest about the size and role of government?

As stated many times, man is a highly social animal, designed primarily for small group, communal living. In ancient human groupings there did need to be some degree of hierarchy and organization. Groups were probably organized around an individual leader who seemed to possess the necessary attributes required during the Paleolithic era. Most group members would have acquiesced to this hierarchy knowing full well that order was preferable to life-threatening disorder.

Would it have been good for a hunter-gatherer clan to have been dominated by a despotic, oppressive, and dictatorial strongman? Probably not. Over time, members of the clan would have most likely grown resentful and maybe even rebellious. Group harmony and cooperation would have suffered.

Groups, in general, tend to thrive better when individuals are free to utilize and express their energy and creativity. Fear and resentment do not invite cooperation, collaboration, or creativity. Communities, both big and small, function more effectively when there is an atmosphere of amiability and trust.

In modern times, democracies do provide enough structure to ensure the delivery of essential governmental services. At the same time, they do not tend to be so oppressive that they stifle creativity and innovation, or precipitate large-scale resentment and rebellion on the part of the citizenry. They are good examples of ungrudging cooperative behavior.

Government can accomplish a lot. In the United States we should credit government with many things such as the interstate highway system, the railways, electricity generation, public education, space exploration

(NASA), healthcare for seniors (Medicare), protection of the environment (EPA), food and drug safety (FDA), and reduction of disease (CDC).

Because these are all very important functions, people are sometimes critical of the government if the performance of these governmental tasks is less than optimal. People get upset if the EPA, FDA, or the CDC isn't performing its very critical task. We should note that nobody writes his or her congressman if the local business is run poorly. But government is held to a higher standard—as well it should.

From Paleolithic times to modern times, some level of group organization has always been beneficial and necessary. Our ancient hunter-gatherer groupings were small and intimate enough for formal government to be unnecessary. Emotions such as shame, fear, remorse, humiliation, affection, compassion, empathy, and sympathy were designed by evolution to facilitate complex group functioning without a formal structure. One might think of these sentiments as a natural form of governmental "checks and balances." However, large, modern, and complex groupings do require a more structured form of governance. But that controlling authority— that government— needs to be mindful of, and in sync with our Paleolithic inheritance and sensibilities.

Let us be fearful of those who thirst for authority, and wary of those that advocate for a government so small it is incapable of exercising its raison d'etre.

Conservatives and Liberals

"Liberals tend to understand that a person can be lucky or unlucky in all matters relevant to his success. Conservatives, however, often make a religious fetish of individualism. - Sam Harris (American author, philosopher, and neuroscientist)

No discussion on politics would be complete without talking about what it means to be a "conservative" or a "liberal" here in the United States.

Starting our discussion in a literal way: to "conserve" is to maintain the status quo. Conservatives generally want to maintain the status quo. To "liberate" is to free something—to change it from its present status to something different. Liberals are generally more open to change. While this literal look is perhaps a good starting point, it leaves much to be desired if we are trying to understand the modern American political landscape.

For example, most people who consider themselves to be "libertarians" also consider themselves to be conservatives. On the other hand, liberals tend to want to conserve the environment. They tend to buy smaller, fuel-efficient cars, and worry more about global warming. These things might seem confusing to the average person. What would a *paleo perspective* be on this matter?

Your Political DNA

Let us start our discussion with the question of whether our political views are influenced by our genetic makeup. Do conservatives and liberals have genetically-based political leanings, or are we born as politically blank slates? Are the brains of conservatives and liberals in some way different? Several psychological and neurological studies seem to indicate that they are indeed.

MRI scans performed on 90 student volunteers at University College London found that self-identified conservatives tended to have a larger amygdala than self-identified liberals (Kanai, Feilden, Firth, Rees, "Political Orientations are Correlated with Brain Structure in Young Adults," *Current Biology*, April 26, 2011). The amygdala concerns itself with the experiencing and processing of emotions such as fear, anxiety, and the "fight or flight" response. Conservatives tend to have a stronger emotional reaction to things that they view as distasteful or upsetting than do liberals.

On the other hand, liberals tended to have a larger anterior cingulate cortex, which concerns itself with the monitoring of uncertainty, and the handling of conflicting information. It mitigates our emotions when we are confronted with unsettling information (affect regulation). And so liberals would tend to be more cognitively reflective when dealing with situations that contain complexity and challenge their long-held worldview.

Conservatives generally respond to threatening situations with more aggression than liberals—who are busy trying to analyze the situation. Most psychologists consider conservatives to be somewhat more suspicious and insecure about the world around them, and therefore desire and strive for more stability. They tend to gravitate toward clear, simple answers to what are arguably complicated problems. Alternatively, liberals are more likely to be comfortable with complexity, nuance, and subtleties, and are more tolerant of an ever-changing world (Jay Dixit, "The Ideological Animal," *Psychology Today,* January 1, 2007).

A case in point would be the belief (or not) in the theory of evolution. According to The Pew Research Center (2013), 67% of Democrats (who tend to be more liberal) believe in human evolution, while only 43% of Republicans (who tend to be more conservative) do. Belief in human evolution requires one to be open to various scientific arguments and scientific complexities. It also may require one to reconsider what one may have been taught as part of one's religious training.

Further evidence of a correlation between genes and political thought can be found in an article by Thomas B. Edsall ("How Much Do Our Genes Influence Our Political Beliefs?" *The New York Times,* July 8, 2014). Here are some excerpts from that article:

> *In "Obedience to Traditional Authority: A heritable factor underlying authoritarianism, conservatism and religiousness," published by the Journal of Personality and Individual Differences in 2013, three psychologists write that "authoritarianism, religiousness and conservatism," which they call the "traditional moral values triad," are "substantially influenced by genetic factors." According to the authors — Steven Ludeke of Colgate, Thomas J. Bouchard of the University of Minnesota, and Wendy Johnson of the University of Edinburgh — all three traits are reflections of "a single, underlying tendency," previously described in one word by Bouchard in a 2006 paper as "traditionalism."*

> *The three psychologists found evidence that they believe demonstrates that authoritarianism, religiosity and conservatism are "different manifestations of a single latent and significantly heritable factor," the tendency to follow conventional authority "in attitudes toward the structure of*

family and society, toward religious conventions, and toward conventional attitudes on political issues."

Working along a parallel path, Amanda Friesen, a political scientist at Indiana University, and Aleksander Ksiazkiewicz, a graduate student in political science at Rice University, concluded from their study (2014) comparing identical and fraternal twins that "the correlation between religious importance and conservatism" is "driven primarily, but usually not exclusively, by genetic factors." For their part, Friesen and Ksiazkiewicz contend that their "findings confirm the existence of a common genetic factor that underlies holding socially conservative policy positions, maintaining traditional values, and placing importance on religion in one's life."

However, when considering the aforementioned studies (and many more like them), let us be mindful of the fact that human behavior is quite complicated. There are many factors, both genetic and environmental, that can affect your political leanings. For instance, whether you live in a rural area or an urban area, your level of education, or your age can influence your political leanings.

To illustrate this point, let us look at some recent data. According to a 2014 Pew Research study, 77% of individuals living in an urban setting describe themselves as liberals, and 75% of individuals living in a rural setting describe themselves as conservatives. This makes sense because living in an urban area presents the individual with more socially complex and ambiguous situations, encouraging and necessitating liberal thought patterns. Rural living is usually less socially complicated, requiring less accommodation of diversity and disorder. We must also keep in mind that individuals with conservative leaning minds might choose to move to a more rural setting, and that individuals with liberal leaning brains might choose to re-locate to an urban setting.

Age is also a factor in political leanings. People over 65 years of age were 5 times more likely to describe themselves as "steadfast" conservatives compared to those under 30 (Pew Research, 2014). This also seems logical because as people age they tend to grow less comfortable with change.

With regard to education, during the recent 2016 election, 52% of individuals with a college degree voted for Clinton, while only 29% voted for Trump. Generally speaking, higher education exposes the individual to more complexity, and helps the individual to analyze and better cope with convoluted technical and social situations. Conservative analysis of (and solution to) social problems are usually simpler and less nuanced.

Let us at least acknowledge that human behavior is complicated, and so politics is likewise complicated.

Individual vs. *Group* - *Present* vs. *Future*

Let us turn our attention in a different direction. *The Paleo Perspective* asks us now to consider two particular human "dispositions"—dispositions that each of us display to some degree or another. As is the case with most human inclinations, one should consider variations in these two dispositions to be a combination of both environmental and genetic factors.

The first so-called disposition involves whether one tends to be more focused on you the *individual*, or one tends to be more focused on the *group*, of which you are a part. In an earlier discussion on morality and altruism, we noted that each of us experience an internal conflict between acting selfishly, to satisfy our own *individual* needs, and acting altruistically, to satisfy the needs of others—the *group*. We should also recall that this internal conflict is a direct result of the fact that evolutionary forces have acted on both the individual and group level.

The second disposition involves whether one tends to be more focused on the *present*, or more on the *future*—more specifically, one's propensity to seek gratification in the present, or one's ability to delay gratification into the future.

Generally speaking, people who are willing to accept discomfort in the here-and-now, do so with the belief that, by doing so, things will be better in the future for both themselves and for others. Generally speaking, those who seek gratification in the present do so with the belief (giving them the benefit of the doubt) that to do so will not necessarily adversely affect their future, or anybody else's future. (Some might argue that this disposition is really about our capacity for denial and our desire to delay pain for as long as possible.)

We will use variations in these two dispositions (*individual* vs. *group* and *present* vs. *future*) to illustrate some fundamental and critical differences between typical conservative thinking and typical liberal thinking. While this perspective is not the only way to analyze political thought, it does provide a calculus from which to operate, and it does correlate well with the previously mentioned neurological and psychological differences between conservatives and liberals.

Interconnected and Coupled

Let us first take note of the fact that these two dispositions are often coupled with one another. We might even consider the *future* vs. *past* disposition to be a function of the *individual* vs. *group* disposition, or that one is a corollary of the other. The most obvious connection is the fact that individuals who are more concerned with the *group* in the *present* also tend to be more concerned with the *group* in the *future*. But there is also the issue of one's willingness to delay gratification in general. A case could be made that self-centeredness seems to make it more difficult to defer gratification. A spoiled child wants what it wants—and it wants it now. It's all part of the same package. Let us look at our hunter-gatherer past to gain some additional insight regarding the connection between these two dispositions.

The first of the dispositions (*individual* vs. *group*) concerns the fact that human beings have evolved with two strong inclinations: the drive to survive as an individual, and the drive to survive as a group member. These two drives are not necessarily at odds with each other. More often than not, our chances of surviving as an individual depend on our group's success. If our group perishes, so do we. And so quite often we put aside selfish behaviors and we cooperate and collaborate with our cohorts. We can certainly imagine our ancestors collaborating and cooperating during a hunt, or coordinating their defenses when under attack from a predator or a rival group. Everybody understands that both in the past and the present, when people act in concert with one another, much more can be accomplished than when individuals go it alone. But cooperation with one's cohorts doesn't always take place in the immediate present.

Sometimes cooperation involves the future as well, and here is where the second disposition comes into play.

The second of these two dispositions (*present* vs. *future*) has to do with the age-old problem of one's willingness to sacrifice some comfort in the here-and-now for the possibility of some benefit in the future. Let us once again look to our past for some guidance.

Complex social cooperation depends heavily on altruistic reciprocity. But reciprocity does not only take place in the present, it also takes place in the future. I may do you a favor now expecting a favor in return at some point in the future.

During ancient times, whoever didn't directly participate in the hunt would have had to trust that when the hunters returned (hopefully with meat), they would share the spoils of the hunt with the rest. Perhaps an individual stayed back in camp or the surrounding area and helped gather roots, berries, or fruit. Perhaps they helped out with childrearing, with keeping the campfire going, or some other camp-related task. Everybody expected that some form of reciprocity would take place when the hunters returned, and everybody expected that the spoils of their labors would be shared.

Cooperative hunting, food gathering, and childrearing required group members to put aside their desire for instant gratification, knowing that cooperating in the present would ensure greater rewards in the future.

The willingness to delay gratification and the inclination to be a cooperative group member are often interdependent social behaviors. They were in the past and they are in the present. Being a helpful, cooperative group member requires one to not only be capable of group-centered altruistic behavior, but it also necessitates that one be willing to delay gratification as part of that cooperative behavior.

Using a Sliding Scale

Man is genetically imbued with many, and sometimes conflicting, instincts. There are times when he behaves selfishly and times when he behaves cooperatively. There are times when man seeks instant gratification and times when he sees value in delaying gratification. There are individuals who in general tend to behave selfishly and tend to seek instant gratification.

There are individuals who in general tend to act altruistically and tend to see value in deferring gratification. Every situation is different, and every individual is unique.

Most people, however, do not exhibit the extremes of these competing predilections. It is more helpful and more accurate to think of an individual's behaviors and inclinations as residing somewhere on a sliding scale, lying somewhere in between the extremes. Of course some people are closer to one extreme than the other.

Let us first proceed by thinking of people as residing somewhere on a sliding scale regarding our first behavioral inclination—where one end of the continuum has its emphasis on the *individual* (self-centered, self-serving behavior), and the other end has its emphasis on the collective—the *group* (altruistic, socially beneficial behavior). We can represent our sliding scale/continuum as follows:

Group -------------------------------------- *Individual*

We shall see by examining various positions taken by liberals and conservatives that liberals tend to be more focused on the *group*, and in general should be positioned further to the left side of the continuum. We shall also see that conservatives tend to be more focused on the *individual*, and should therefore be positioned further to the right side of the continuum. This is one of the critical distinctions between typical liberal and typical conservative thinking. We can express our sliding scale/continuum as follows:

Group (liberals) ------------------- *Individual (conservatives)*

Moving on to the second behavioral inclination, whether one focuses on the *present* (short term) or on the *future* (long term), we shall observe that conservatives are more likely to be focused on the short/shorter term, and liberals are more likely to be focused on the long/longer term. This is another key distinction between typical liberal thinking and typical conservative thinking. We can express our sliding scale/continuum as follows:

Future (liberals) ---------------------*Present (conservatives)*

We have also previously stated that these two human traits (dispositions) are not mutually exclusive, that there is a coupling between focusing on the *group* instead of the *individual*, and focusing on the *future* instead of the immediate *present*. We can represent the combined sliding scales as follows:

Group/Future (liberals)-----------------Individual/Present (conservatives)

Let us use this sliding scale—this political spectrum—to talk about how these philosophical differences influence how liberals and conservatives view some current issues. We shall begin by furthering our description of stereotypical conservative thinking and stereotypical liberal thinking, and then examine some areas of disagreement between liberals and conservatives in America today.

As we talk about conservatives and liberals, keep the concept of the sliding scale in mind. Some conservatives are at the extreme right side of the scale, but others who self identify as conservatives might be a bit closer to the center—center right. Likewise, not all the people who self identify as liberals belong on the extreme left side of the sliding scale, but may be a bit closer to the center—center left. The concept of a political sliding scale (spectrum) acknowledges this aspect of human nature.

Typical Conservative Thinking

"By the grace of reality and the nature of life, man—every man—is an end in himself, he exists for his own sake, and the achievement of his own happiness is his highest moral purpose" - Ayn Rand (Russian-American novelist and philosopher)

We have stated that conservatives tend to gravitate toward the individual side (right side) of the political spectrum. They believe that their rights, the rights of the individual, are paramount, and that a person should be free to live out their life with as little interference from other people, or from society, as possible. Their focus is not on the collective or society as a whole. Their focus is on their rights and their well-being as individuals. They tend to believe that their well-being, success, and even their survival is a direct result of their own efforts and initiative. A conservative might even feel that they are better off on their own—that other people (society) might drag them down and be detrimental to their well-being.

This disposition correlates very well with the aforementioned evidence of a slightly enlarged amygdala in the conservative brain. All groups are heterogeneous to a certain extent because they are made up of unique individuals—individuals with different needs, wants, and opinions. More often than not, groups embody some degree of diversity, and with diversity comes the potential for disagreement and conflict. Conservatives do not do as well with conflict and contradictory information, or any challenge to their well-established sense of order. Dealing with inherent group heterogeneity can be unsettling to the conservative mindset. Focusing just on oneself can help give the conservative mind an enhanced sense of control. (There is some irony in the fact that when conservatives, as individuals, feel that they are losing control over a situation, they sometimes form or join a group of like-minded individuals in an effort to re-establish a sense of control.)

Conservatives also tend to be more focused on the here-and-now, and less focused on possible future concerns. This is why we have placed them further to the right on our *future* vs. *present* scale. For example, when it comes to business or economic issues, they tend to favor a freer, unfettered market, and are generally less concerned with the potential social and environmental effects that such policies might precipitate in the future. They would argue that it is more important to encourage economic growth in the here-and-now than to be concerned with any possible future environmental damage that may or may not result.

Here too we can see the conservative's aversion to disorder and change. The future always carries with it the possibility of turmoil and unrest—change. Worrying about the future often requires the conservative mind to deal with uncertainty, variability, and complexity. Contending with a potentially threatening future tends to push the conservative out of his comfort zone.

Most conservatives advocate for smaller government, fewer laws, and fewer regulations. Laws and regulations (government) impinge upon an individual's right to exercise their free will. For instance, they might argue against governmental restrictions on private gun ownership. Similarly, conservatives often object to the government interfering with personal choices such as various municipal smoking prohibitions, mandated calorie displays on menus, and restrictions on the size of sugary drinks. The typical conservative would argue that it is up to each individual to make decisions on such matters without governmental interference.

Governmental rules and regulations all represent the subjugation of the individual to the will of the *group*. For the conservative, acquiescing to the *group's* will represents a significant erosion of self-determination, a loss of freedom, and a loss of control—all of which are antithetical to the conservative mind.

Typical Liberal Thinking

"I believe in a relatively equal society, supported by institutions that limit extremes of wealth and poverty. I believe in democracy, civil liberties, and the rule of law. That makes me a liberal, and I'm proud of it."
- Paul Krugman (American economist and columnist)

We have stated that liberals gravitate to the left side of the political spectrum because they tend to focus more on the *group* (collective). In general, they believe that everyone is better off when we focus on the well-being of the collective. Because of this emphasis on the *group*, most liberals advocate for a larger, more active, and more involved government. They see laws and regulations as being necessary and essential for the orderly operation and success of any society. They feel that without these laws, regulations, and guidelines pandemonium and chaos would result, and everyone would suffer.

Liberals would concede that governmental controls do impinge upon an individual's exercise of their free will, but that this is necessary and a relatively small price to pay for an orderly, compassionate society—one that ensures the well-being of all its constituents. For example, liberals would probably agree with municipal regulations regarding smoking in a public space, product safety labeling, limits on the size of soft drinks, and anything else that would contribute to overall public health. Progressives (liberals) see these trade-offs between individual liberty and greater social good as being well worth the concession.

Here we see the liberal's tolerance for dealing with the inherent complexity, and the necessary compromise, that come with group living.

And so the liberal is more comfortable with living within the socially confounding and demanding collective.

Liberals also tend to be more focused on the future. They are more likely to be willing to delay gratification in an effort to avoid possible future problems, and to also potentially reap some future benefits. For instance, when it comes to business and economic issues, they tend to see value in significant governmental regulation of markets. They would argue that unregulated businesses would put today's profits ahead of any social or environmental damage that might result in the future.

What we observe with the liberal's focus on the future is indicative of his capacity and willingness to deal with the uncertainty and the complexity that the future often represents. And so the liberal is more comfortable being mindful and focused on the future—whatever it may bring.

Let us now look at some particular hot button issues on the American political scene today. As we do, let us keep in mind which side of our two sliding scales (*group* vs. *individual* and *future* vs. *present*) that conservatives and liberals each gravitate towards. Let us also be mindful of the conservative's aversion to conflicting and contradictory information—hence the appeal of simple solutions. Likewise, let us note the liberal's comfort and proclivity to entertain discord and complexity—hence the toleration of complex and sometimes paradoxical solutions.

Example 1: Global Warming and the Environment

The stereotypical conservative would most likely resent the government's efforts to increase automobile fuel efficiency, and would advocate for the individual's right to drive any kind of automobile they desired, no matter what the fuel efficiency happens to be. Governmental fuel efficiency regulations represent a classic case of the individual relinquishing control to the collective.

Conservatives, in general, tend to resent efforts by the government to limit the burning of fossil fuels and usually favor more coal, oil, and natural gas exploration. Demonstrating a reluctance to entertain potential complications, they would focus less on possible future environmental damage and more on satisfying more tangible and immediate needs and desires. You may recall that the stereotypical conservative is further to the right on our sliding scale—more towards the individual and immediate gratification.

Stereotypical liberals tend to favor governmentally prescribed clean air standards and fuel efficiency targets. They are more open to accepting the inconvenient facts, concepts, and theories that the belief in human-based climate change requires. Consequently, the liberal is more likely to accept efforts to curb or tax the use of fossil fuels.

For the liberal, the well-being of the collective is more important than satisfying the desires of the individual. This position might be an inconvenience in the present (higher energy costs), but the potential benefits to society and the Earth is well worth the present day sacrifices. You may recall that the stereotypical liberal is further to the left on our sliding scale—more towards the group and delayed gratification.

A similar case in point would be fracking. Most conservatives are in favor of fracking. Fracking increases the supply of fossil fuels—giving the individual greater freedom of choice—and brings down the cost of energy in the short term.

Most liberals are suspicious of any future deleterious environmental effects that fracking might cause. They are more willing to pay for higher energy costs in the present to safeguard the future. As with global warming, entertaining the scientific pros and cons of fracking requires a tolerance for sometimes-conflicting technical details, and a willingness to confront "inconvenient truths."

Example 2: Taxes

Most conservatives advocate for lower taxes. They argue that individuals know best how to manage and spend their own money, and that the government does not have the right to take away their money and redistribute it to other people and other parts of society (*group*).

Similarly, many conservatives are leery of some of the government's safety net programs like food stamps, unemployment insurance, or government-sponsored health insurance. They argue that it is each individual's responsibility to support themselves and to take care of their own health needs. It is not the individual's responsibility to help support and take care of the rest of society. Once again, we see here the conservative's reluctance to relinquish a measure of individual control to the collective.

What do we expect a liberal's view on taxes to be? Most taxes are used by the government to provide public services. These services include things like public education, emergency medical services, road maintenance, sewage disposal, food stamps, and unemployment insurance. On his own, an individual cannot provide for such services. These are services that are best provided by a governmental agency, and are services that benefit all of society. Most progressives (liberals) do see taxes as an encumbrance, but one that is well worth the cost. This position is consistent with the liberal's tolerance for the compromises that complex societies (groups) require.

As previously pointed out, urban areas tend to be more liberal than rural areas. This seems reasonable because people living in densely populated areas have more need of public services like garbage collection, mass transit, and sewage disposal. Quite naturally, people living out in rural farmlands see less of a need for tax funded "unnecessary" public works.

Example 3: Public Unions and Public Pensions

The stereotypical conservative is not a big supporter of unions—especially public employee unions. Unions represent people forming a group to pursue some sort of collective benefit. A union's primary purpose is to enhance the well-being of its members, be it salary or working conditions. This kind of group behavior runs contrary to the conservatives focus on the individual's right and responsibility to act on his own behalf and advocate for himself. Conservatives tend to resent the fact that many times union membership is mandatory, and that union membership requires that each member give up a degree of autonomy to the union (group).

Public employee pensions, most likely resulting from a union's collective bargaining efforts, can be particularly irritating to the conservative. These pensions are funded by taxes paid by any individual who resides within the municipality served by the public employee. As previously stated, most conservatives resent the mandatory transfer of money from the individual to the collective at large. This represents the subjugation of the individual by the collective, and the loss of a measure of individual autonomy.

The stereotypical liberal is often a supporter of unions, both public and private. They view the formation of unions as a good example of people sticking together to enhance the well-being of the group. The liberal does understand that to enjoy the benefits of union membership, one must be willing to follow union rules and regulations. But once again, we see willingness on the part of the liberal to tolerate the complexity and sacrifice that the group experience requires. Union members must be willing to give up a measure of individual autonomy for the greater good of the collective.

With regard to public employee pensions—and the taxes that must be collected to pay for those pensions—liberals tend to see value in paying for the services that those pensions represent. For them, this reasonable degree of financial subjugation is well worth the resulting public good that public services bring to a community.

Example 4: Social and Economic Inequality

In general, the conservative will tolerate social and economic inequality more than the liberal. To the conservative, inequality is a result of natural variations in the motivation, talent, and capabilities of the individual. We can perhaps think of this as a type of social Darwinism, whereby those who are more fit reap the rewards of their superiority. Conservatives might also argue that significant inequality is good in that it provides motivation for people to work harder and be more productive. The emphasis here is on individual freedom, initiative, and responsibility.

The liberal tends to see significant socioeconomic inequality as being problematic. They argue that large-scale inequality is not so much a function of significant differences in individual talent, but instead a result of a system that favors the rich and the well-connected. It undermines the overall health of the collective by encouraging class warfare and diminishing the necessary comity and goodwill for the group to function as a cohesive whole.

Example 5: The Federal Minimum Wage

The stereotypical conservative most likely does not favor the federal government setting a minimum wage. They reason that each of us, as individuals, are responsible for getting a well-paying job and securing our own financial future. They argue that it is up to the interaction of the autonomous individual worker and the independent business owner—operating in a free market—to determine wages, and that any action by the government (collective) is an infringement on the rights of both the individual worker and the business owner. The conservative feels that artificial wage minimums upset free-market wage-price machinations and curtails the business owner's right to make as much of a profit as they desire. A federal minimum wage represents the individual giving up control to the collective.

Stereotypical liberals are more likely to be in favor of setting minimum wage standards. They argue that ensuring a living wage would benefit society as a whole by reducing things like poverty, crime, and drug addiction. The liberal would agree that when the government sets a minimum wage, individuals are relinquishing a small measure of their freedoms, but the sacrifice is well worth the overall social benefit.

Example 6: Voter Identification Laws

The stereotypical conservative would most likely focus on the individual's responsibility to have the necessary official identification documents that would establish one's right to vote. They feel comfortable with the social order that verification brings, and would be leery of any governmental policies that might give special accommodations to select groups of people. The emphasis here is on individual responsibilities and order.

The stereotypical liberal would be concerned that the poor or racial minorities might not have access to official birth or residence documentation. These complications and inconveniences exemplify a lack of social order—irritating to the conservative, but well-tolerated by the liberal. In advocating for the poor or for minorities, liberals reason that society as a whole benefits when everyone participates in the democratic process and everyone's views are represented. The emphasis here is on creating a healthy society, one characterized by inclusion, equality, and fairness. Everyone is better off if the group is better off.

Example 7: Civil Rights

In general, stereotypical conservatives are leery of sacrificing their individual civil rights to accommodate the needs of others—like gay people, transgender people, or racial minorities. The conservative tends to see things like gay marriage, gay participation in the military, or gay rights to be an assault on their individual right to have a dissenting opinion on the morality of homosexual behavior. They view the issue of gay rights as an assault on their religious freedom.

For example, a conservative baker or photographer might refuse to do business with a gay couple. The conservative would argue that the community at large, or the government, should not force its opinion on the individual, possibly violating that individual's religious sensibilities. The acceptance of gay marriage also represents a move away from the historical status quo, and as previously mentioned, conservatives tend to resist notable social change.

Another example regarding civil rights might be a conservative's objection to affirmative action in college admission, or racial considerations of any kind. They argue that any special consideration given to a protected class (group) violates the rights of any individual who does not happen to be a member of that protected class. This too represents acquiescence to the group's will and a loss of individual freedom. For conservatives, the emphasis is on individual rights rather than on group rights.

The liberal looks at the issue of gay rights not only in terms of the rights of gay people as individuals, but also in terms of their rights as group members. More attention is paid to how society as a whole should function, with the belief that everybody benefits when the collective acts with moral rectitude and tolerance towards all. The emphasis here is on group well-being. In general, liberalism reflects a tolerance for diversity within the collective for the sake of the health of the collective. Minority rights are group rights.

With respect to racial quotas, or any other program that attempts to enhance the well-being of minorities, liberals argue that these kinds of programs and policies are an attempt to level the playing field for minorities, and are necessary to insure a more equitable society in general. Liberals are more willing to sacrifice some of the rights they enjoy as individuals to accommodate the needs of various sub-groups—such as women, gay people, or racial minorities. They more easily accept the fact that living in a heterogeneous grouping requires compromise, and that individuals must be willing to forgo a measure of their rights for the benefit of society as a whole.

Liberals feel that gay rights are civil rights. Black rights are civil rights. Women's rights are civil rights. Guaranteeing the civil rights of any minority helps ensure the well-being of the entire collective.

Example 8: Abortion and Contraception

This is perhaps the most difficult one to explain. Most people who consider themselves to be conservatives are usually pro-life (anti-abortion), and most people who consider themselves to be liberals are usually in favor of a woman's right to choose. At first glance, the issue of legalized abortion might seem to contradict the stereotypical perspectives of liberals and conservatives that is being presented here. After all, wouldn't a conservative be concerned with the individual's right to decide for herself whether or not to have an abortion? Wouldn't the liberal feel that the government (the collective) has a right to exercise its will over the individual and prohibit abortions for the overall well-being of society?

However, the stereotypical conservative tends to see the abortion issue as a moral one—a religious one—where the termination of a pregnancy results in the death of an unborn human being. As previously mentioned, conservatives tend to be more religious than liberals. In the case of abortion, they see themselves as a protector of the unborn (individual) from the secular leanings of the masses (group). The conservative is also more likely to view this issue in "black and white," arguing that the fetus is fully human upon conception.

Similarly, conservatives also tend to view the issue of readily available contraception in religious or moral terms. They tend to feel that by making contraception easily available we are inviting the young and the unmarried to engage in sexually promiscuous behaviors. Once again, the conservative sees the necessity to protect the individual from the secular leanings of society at large. This perspective seems to express less concern with any possible future sociological consequences of unplanned births and single parenthood, and instead seems more concerned with the moral and religious dimensions of unmarried sexual activity.

To make sense of what on the surface might appear to be a reversal of liberal and conservative values, we need to consider the idea that conservatives tend to see a person's religious beliefs as part of their rights as individuals. They tend to see legalized abortions and readily available contraception as examples of the state imposing the secular values of the collective on the individual. The stereotypical conservative is resistant to the idea that an individual must subjugate his or her religious beliefs to the will of the collective—even if the will of the collective represents a majority opinion. This is an example of individual rights over group consensus.

The stereotypical liberal tends to see the abortion issue as group rights (women's) versus the religious sensibilities of the individual. An individual's religious beliefs should not universally trump the will of the collective. The liberal's overarching perspective is that society as a whole is better off when all the various segments and groupings (women) within that society enjoy the full measure of the freedoms offered by that society. Women's rights are civil rights. Once again, the emphasis is put on the overall health of society (the group), not on the absolute right of the individual.

One of the sticking points in the abortion conundrum is the status of the unborn. As previously mentioned, the conservative tends to view the fetus as having total and absolute human status. Also previously mentioned is the fact that the liberal mind is more comfortable with nuance, and is willing to see matters in shades of gray rather than in black or white. As such, he is disinclined to draw a distinct line to determine whether or not—or at what point—should full human status be granted to the fetus.

Often, when discussing any issue involving religious beliefs, the question of the separation of church and state comes into play. The stereotypical liberal sees the concept of the separation of church and state as an example of a compromise that individuals must make in order to ensure the smooth and efficacious workings of a multi-religious society.

The conservative sees it as a case of individual rights being trampled by the collective—a loss of individual freedom.

The Eye of the Beholder

A liberal might consider some of the typical positions and attitudes of the conservative to be selfish, and that the conservative is too focused on himself and ignores the needs of society around him. Similarly, in the liberal's eyes, the conservative is also too concerned with satisfying his present day needs and desires (like unbridled fossil fuel consumption) and is unwilling to postpone gratification to ensure a better future for all.

A conservative might consider some of the typical positions and attitudes of the liberal to be naïve and utopian—unrealistic to say the least. To the conservative eye the liberal is too focused on the collective, ignoring the needs and rights of the individual, and thereby subjecting the will of the individual to the will of the collective. Similarly, in the conservative's eye, the liberal is overly concerned with possible future deleterious events (like global warming)—events that the conservative doubts might ever come to pass.

Right-Wing versus Left-Wing

Conservative thinking is sometimes referred to as "right-wing" thinking and liberal thinking is sometimes referred to as "left-wing" thinking. If we look at *right wing* versus *left wing* politics on our *paleo perspective* sliding scale, *left-wingers* tend to be further to the left side of the continuum, and *right-wingers* tend to be further to the right side of the continuum. Some might even say that extreme *right-wing* politics starts to resemble fascism. And some might say that extreme *left-wing* politics starts to resemble socialism or communism. (We might note at this point that any form or degree of government is a form of socialism.)

Historically, fascism has been associated with some ethnic and/or racial issues. Nazi ideology included the idea of a superior "Aryan" race, which excluded people of Jewish and African descent. And so the concept of "us" to a fascist might only include the fascist's ethnic and racial cohorts.

Socialism, on the other hand, tends to be more inclusive—with a bigger "us"—that ideally includes all ethnic and racial variations.

Man evolved in the group setting, and so there has always been an "us." The question is how inclusive is the "us." Does it include multiple ethnic, racial, and religious groupings, or is it more selective?

Who is Correct?

So, who is correct: conservatives or liberals? One might assume that there isn't a correct position on such matters, or that the truth lies somewhere in the middle—between the conservative and the liberal positions. To assume that the truth *always* lies somewhere in the middle is foolish. There are such things as facts. Human evolution is either a fact or not. Human induced global warming is either a fact or not. The vast majority of scientists around the world have concluded that global warming and evolution are scientifically verifiable facts. One's political leanings do not alter these facts.

"Everyone is entitled to his own opinion, but not his own facts."
- Patrick Moynihan (American senator and sociologist)

What would a *paleo perspective* on such matters be? Homo sapiens were designed to function best in a communal type of setting. In such a setting it was wise to focus on the well-being of the collective. *The Paleo Perspective* points out that man is in fact quite tribal, and his survival depended on his ability and willingness to cooperate with his fellow humans.

Author Ayn Rand has been a significant influence on American conservative and libertarian thought. She is quoted as saying: "If any civilization is to survive, it is the morality of altruism that men have to reject." *The Paleo Perspective* couldn't disagree more. It is precisely because of our morality-based altruism, our cooperation and collaboration, that we have been so successful as a species.

Ironically, the conservative's exaltation of the individual, with its illusion of self-sufficiency, is only possible because the social fabric that supports and sustains us as individuals is so complex, ubiquitous, and effective. It is easily taken for granted. With the possible exceptions of a few isolated, "off the grid" individuals, the modern individual flourishes in the context of the social milieu that surrounds and supports him—the social "nest" if you will.

> *"There is no such thing as a self-made man. You will reach your goals only with the help of others." - George Shinn (American author and sports team owner)*

A case in point is the fact that many times a single individual gets credit for an invention or a major advancement in science: Alexander Graham Bell's telephone, Thomas Edison's light bulb, James Watson's discovery of DNA, or Albert Einstein's theory of relativity. All of these individuals would be the first to admit that they alone should not be given total credit for their accomplishments, that their work was in conjunction with the work of other scientists of the day, and that their discovery would not have been possible without the knowledge and contributions of those around them and all those who came before them.

Consider the modern computer, cell phone, or automobile. None of these devices would be useful or even possible without the technical, logistical, and social fabric that supports their existence (cable wiring, cell towers, roads, bridges, traffic rules, etc.). They simply would not exist without a rich and highly complex social community—the group.

A Cautionary Note

One wonders if perhaps the biggest threat to civilization is the rise of the individual—the obsession with the self. Many times an individual will focus only on himself, look to satisfy his personal desires, and seek to maximize the accumulation of worldly goods. Are we operating under the delusion that at this point in time—the population density and heterogeneity being what it is—we can operate as detached individuals instead of cooperative group members? Are we losing our will for altruistic reciprocity and delayed gratification—the very things that made us a successful species from the start?

When the group (society) is large, impersonal, and complex, individuals can lose sight of the fact that they are part of a group and have a responsibility toward that group. It is foolish to ignore the successes of cooperative living. Our species would not have survived if the individual were considered to be the center of the universe. If modern society is to move in a positive direction, there is the need for more community—not less.

Liberal and Conservative States

It is often said that our states (within the USA) are laboratories for public policy—examples from which to see what works and what doesn't. Let us now have a look at some side-by-side comparisons of conservative and liberal principles in practice, state-by-state. Let us use the following nomenclature for our comparisons:

Liberal	versus	*Conservative*
Left	versus	*Right*
Democrat	versus	*Republican*
Blue States	versus	*Red States*

Having given some examples of both typical conservative and typical liberal thinking, let us now investigate how each of these two perspectives plays out across the current American landscape. We will look at several socioeconomic variables, comparing conservative-leaning states with liberal-leaning states. The data will sample lifestyle, health, education, and economic issues. Most of the results will be in the format of a "red state" versus "blue state" comparison. The assumption being that Republican-leaning (red) states are generally more conservative than Democratic-leaning (blue) states. We can also think of "purple" states as being somewhere in the middle—perhaps voting Democratic one year and Republican the next. We should also note that throughout history, various political parties have been conservative at one point in time and more progressive at other points in time. For instance, the Republican Party of Abraham Lincoln was much more liberal than the Republican Party of today. The southern faction of the Democratic Party during the fifties and sixties was much more conservative than the current Democratic Party.

The following comparisons are displayed as the 10 states with the "lowest rates" versus the 10 states with the "highest rates" within each of the socioeconomic categories. The color designations (blue, red or purple) used here are a summary of the 2000, 2004, 2008, 2012, and 2016 presidential elections.

Median Household Income

Highest	*Lowest*
Maryland – blue	Mississippi – red
New Hampshire – blue	West Virginia – red
Hawaii – blue	Alabama – red
Connecticut – blue	Louisiana – red
Alaska – red	Kentucky – red
Minnesota – blue	Tennessee – red
Virginia – purple	Arkansas – red
New Jersey – blue	South Carolina – red
Utah – red	Florida – purple
Massachusetts – blue	New Mexico – blue

We can see that 7 out of 10 states with the highest household incomes are liberal-leaning, and 8 out of 10 states with the lowest household incomes are conservative-leaning. (US Census Bureau, 2014)

Households Below Federal Poverty Line

Lowest percentage	*Highest percentage*
New Hampshire – blue	Mississippi – red
Maryland – blue	Louisiana – red
Oregon – blue	New Mexico – blue
Minnesota – blue	Alabama – red
Hawaii – blue	Texas – red

Delaware – blue

Utah – red

Virginia – purple

Nebraska - red

Connecticut - blue

Arkansas – red

Oklahoma – red

West Virginia – red

Arizona – red

South Carolina – red

We can see that 7 out of 10 states with the lowest percentage of households below the poverty line are liberal-leaning, and 9 out of 10 states with the highest percent of households below the poverty line are conservative-leaning. (US Census Bureau, 2014)

Children Below the Federal Poverty Line

Lowest percentage

Highest percentage

North Dakota – red

Maryland – blue

Alaska – red

Minnesota – blue

Connecticut – blue

Utah – red

Virginia – purple

Massachusetts – blue

New Jersey – blue

Vermont – blue

Mississippi – red

New Mexico – blue

Arkansas – red

Louisiana – red

Alabama – red

Georgia – red

Arizona – red

South Carolina – red

Kentucky – red

North Carolina – purple

We can see that 6 out of 10 states with the lowest percent of children below the poverty line are liberal-leaning, and 8 out of 10 states with the highest percent of children below the poverty line are conservative-leaning. (Carsey Institute, University of New Hampshire, 2013)

Dependency on Federal Funding

Least dependent	*Most dependent*
New Jersey – blue	New Mexico – blue
Delaware – blue	Mississippi – red
Illinois – blue	Kentucky – red
Minnesota – blue	Alabama – red
Kansas – red	Montana – red
California – blue	West Virginia – red
Connecticut – blue	Arizona – red
Massachusetts – blue	Louisiana – red
Nebraska – red	South Dakota – red
Ohio – purple	Maine – blue

We can see that 7 out of 10 states least dependent on federal funds are liberal-leaning, and 8 out of 10 states most dependent on federal funds are conservative-leaning. (WalletHub, 2015)

Bachelor's Degree

Highest percentage	*Lowest percentage*
Massachusetts – blue	West Virginia – red
Colorado – purple	Arkansas – red
Maryland – blue	Mississippi – red
Connecticut – blue	Kentucky – red
New Jersey – blue	Louisiana – red
Virginia – purple	Nevada – purple
Vermont – blue	Alabama – red
New Hampshire – blue	Indiana – red
New York – blue	Oklahoma – red
Minnesota – blue	Tennessee – red

We can see that 8 out of 10 states with the highest percentage of people with a bachelor's degree are liberal-leaning, and 9 out of 10 states with the lowest percentage of people with a bachelor's degree are conservative-leaning. (US Census Bureau, 2011 - 2015)

Life Expectancy

Longest	*Shortest*
Hawaii – blue	Mississippi – red
Minnesota – blue	West Virginia - red
Connecticut – blue	Alabama – red
California – blue	Louisiana – red
Massachusetts – blue	Oklahoma – red
New York – blue	Arkansas – red
Vermont – blue	Kentucky – red
New Hampshire – blue	Tennessee – red
New Jersey – blue	South Carolina – red
Utah – red	Georgia – red

We can see that 9 out of 10 states with the longest life expectancy are liberal-leaning, and all 10 states with the shortest life expectancy are conservative-leaning. (Center for Disease Control, 2009)

Smoking

Lowest rates	*Highest rates*
Utah – red	Kentucky – red
California – blue	West Virginia - red
Minnesota – blue	Mississippi - red
Massachusetts – blue	Oklahoma - red
New Jersey – blue	Ohio - purple
Maryland –blue	Missouri - red

Washington – blue	Indiana - red
Rhode Island – blue	Louisiana - red
Colorado – blue	Tennessee - red
Arizona – red	Michigan – blue

We can see that 8 out of the 10 states with the lowest rate of smokers are liberal-leaning, and that 8 out of 10 states with the highest rate of smoking are conservative-leaning. (US Gallup Poll, 2014)

Obesity

Least obese	*Most obese*
Hawaii – blue	Arkansas – red
Massachusetts – blue	West Virginia – red
California – blue	Mississippi – red
Vermont – blue	Louisiana – red
Utah – red	Alabama – red
Florida – purple	Oklahoma – red
Connecticut – blue	Indiana – red
Montana – red	Ohio – purple
New Jersey – blue	North Dakota – red
Rhode Island – blue	South Carolina – red

We can see that 7 out of 10 of the states with the lowest obesity rates are liberal-leaning, and 9 out of 10 states with the highest levels of obesity are conservative-leaning. (Robert Woods Johnson Foundation, 2014)

Health Care

Best healthcare	*Worst healthcare*
Hawaii – blue	Louisiana – red
Vermont – blue	Mississippi – red

Maine – blue	Arkansas – red
Minnesota – blue	West Virginia – red
New Hampshire – blue	Alabama – red
Connecticut – blue	Oklahoma – red
Utah – red	Kentucky – red
Colorado – purple	Tennessee – red
Washington – blue	South Carolina – red
Nebraska – red	Indiana – red

We can see that 7 out of 10 states with the best healthcare are liberal-leaning, and all 10 of the states with the worst healthcare are conservative-leaning. (United Health Foundation's Annual Report, 2015)

Divorce Rates

Lowest rates	*Highest rates*
Massachusetts – blue	Tennessee – red
Maryland – blue	New Mexico – purple
Illinois – blue	Kentucky – red
California – blue	West Virginia – red
North Dakota – red	Arkansas – red
Utah – red	Oregon – blue
Pennsylvania – purple	Oklahoma – red
Hawaii – blue	Maine – blue
New York – blue	Florida – purple
New Jersey – blue	Nevada – purple

We can see that 7 out of 10 states with the lowest divorce rates are liberal-leaning, and 7 out of 10 states with the highest divorce rates are conservative-leaning (or purple). (US Census, 2012)

Suicide Rates

Lowest Rates	*Highest Rates*
New Jersey – blue	Wyoming – red
New York – blue	Alaska – red
Massachusetts – blue	Montana - red
Rhode Island – blue	New Mexico – blue
Maryland – blue	Utah – red
Illinois – blue	Colorado – purple
Connecticut – blue	Idaho – red
California – blue	Nevada – purple
Georgia – red	Oregon – blue
Texas – red	Oklahoma - red

We can see that 8 out of 10 states the states with the lowest rates of suicide are liberal-leaning, and 6 out of 10 states with highest suicide rates are conservative-leaning. (Center for Disease Control, 2014)

Teen Pregnancy

Lowest rates	*Highest rates*
New Hampshire – blue	Mississippi – red
Vermont – blue	Texas – red
Minnesota – blue	Arkansas – red
Massachusetts – blue	Louisiana – red
Utah – red	Oklahoma – red
Wisconsin – purple	Nevada – purple
North Dakota – red	Delaware – blue
Nebraska – red	South Carolina – red
Iowa – purple	Hawaii – blue
Rhode Island – blue	Georgia – red

We can see that 5 out of 10 states with the lowest rate of teen pregnancy are liberal-leaning states, and that 7 out of 10 states with the highest rate of teen pregnancy are conservative-leaning states. (Guttmacher Institute, 2010)

Teen Birth Rate

Lowest rates	*Highest rates*
Massachusetts – blue	Alabama – red
New Hampshire – blue	Arkansas – red
Connecticut – blue	Kentucky – red
New Jersey – blue	Louisiana – red
Vermont – blue	Oklahoma – red
Minnesota – blue	Mississippi – red
Rhode Island – blue	New Mexico – purple
New York – blue	Tennessee – red
Maine – blue	West Virginia – red
Maryland – blue	Texas – red

We can see that 10 out of 10 of the states with the lowest teen birth rates are liberal-leaning, and 9 out of 10 states with the highest teen birth rates are conservative-leaning. (National Campaign to Prevent Teen and Unwanted Pregnancy, 2014)

Youth Risk

Some additional data was recently published in *The New York Times* (Nicholas Kristof, "Blue States Practice the Family Values Red States Preach," Nov 18, 2017) regarding the *Youth Risk Behavior Survey*. According to the survey (32 states), 4 out of 5 of the states with the highest percentage of high school students who say they have had sex voted Republican in 2016, while 4 out of 5 of the states with the lowest percentage voted Democratic.

Some Conclusions

It is quite striking how different liberal-leaning and conservative-leaning states are in these state-by-state comparisons. Do these results make sense? What, if anything, is this data telling us?

One of the key concepts in *The Paleo Perspective* is the interplay between man's instinct to act as a self-serving individual and his instinct to be a cooperative group member. The fundamental question being looked at with these various blue state – red state comparisons is whether we are better off when our focus is more on individual liberty or when our focus is more on group cooperation. Which emphasis results in better sociological outcomes? From the above comparisons it is apparent that for some very important sociological issues/metrics, liberal-leaning states fare better than conservative-leaning states. The side-by-side comparisons are fairly consistent and convincing.

From a *paleo perspective,* these results are not surprising. After all, for the better part of the last 200,000 years, man's success has depended on his instinct to cooperate with his fellow human beings. It should come as no surprise that when we move away from socially cooperative behavior patterns—like investing in public education, universal healthcare, and other social services—we suffer both as individuals and as a group.

To be fair, we should also note that even though blue-designated states are liberal-leaning, they are not entirely populated by people at the extreme left hand side of the political spectrum. Similarly, even though red-designated states are conservative-leaning, they are not entirely populated with individuals at the extreme right side of the political spectrum.

However, the statistics presented in these various state-by-state comparisons should at least give us pause for thought. Too much emphasis on the individual, and too little consideration for the collective, does appear to have its socioeconomic drawbacks.

The United States happens to be somewhat more conservative than many of the countries of Europe. Perhaps it has to do with the fact that America was part of a newly-discovered continent—a frontier where there were no laws or government to speak of. These pioneers were mostly on their own, with little social structure to depend on. Survival was very much an individual endeavor. It is understandable how this young country might value the self-sufficient, liberty-loving individual, as most conservatives do.

Be that as it may, what would we find if we were to compare the relatively conservative United States with the rest of the world in terms of various socio-economic criteria?

Liberal and Conservative Countries

The *Social Progress Index* measures the extent to which over 100 countries throughout the world provide for the social and environmental needs of their citizens. Fifty-four indicators such as nutrition, shelter, safety, health, wellness, human rights, access to information, and education are compared for these countries. (The index is published by the nonprofit *Social Progress Imperative* and is based on the writings of Amartya Sen, Douglass North, and Joseph Stiglitz.) How does the United States, which is a relatively conservative-leaning country, compare with the rest of the world?

According to the 2014 *Social Progress Index,* the United States ranks number 16—behind Norway, Sweden, Switzerland, Iceland, New Zealand, Canada, Finland, Denmark, Netherlands, Australia, United Kingdom, Austria, Germany, and Japan. We should note that most of the countries that placed higher than the United States are considered to be more liberal in terms of their more generous government-sponsored programs and their more comprehensive social safety net.

We can also look at how the United States compares to other nations around the world when it comes to income inequality. Income inequality is often expressed in terms of something called the *Gini Index.* The *Gini Index* is a summary statistic that measures the dispersion of incomes on a scale of zero to one. Zero means everyone has exactly the same income, and one means one person has all the income.

According to data from the *Organization for Economic Cooperation and Development* (OECD), the U.S. has one of the most unequal income distributions in the developed world, even after taxes and social-welfare policies are taken into account. The United States ranked 27th out of 31 OECD countries for the year 2010.

Another indication of how the conservative-leaning United States compares with other countries around the world is *The Quality of Life Index*. This index is generated by *Numbeo*, which is the world's largest database of user-contributed data about cities and countries worldwide. *Numbeo* provides current and timely information on world living conditions—including cost of living, housing, health care, traffic, crime, and pollution. Data for the year 2016 indicates that the United States was 12th—behind Switzerland, Denmark, New Zealand, Germany, Australia, Austria, Netherlands, Norway, Sweden, and Finland. Most of the countries that ranked higher than the United States are considered to have more liberal social policies than the USA.

There are of course many reasons why some countries stack up better than others when it comes to these comparisons. Comparing countries with different climates, different geographies, and access to different natural resources is problematic to say the least. However, some of these comparisons should give us pause for thought regarding the efficacy of some American conservative-leaning policies. At the very least it should make us cautious of extreme *right wing* politics. More often than not, extreme policies of any stripe can spell trouble.

Competition or Cooperation ("Me or We")

Each of us considers ourselves to be a unique individual, but we must also not forget or deny our common evolutionary design. Ultimately, everything is about survival—both self-centered behavior and altruistic cooperation. But cooperation was unquestionably crucial in the Paleolithic hunter-gatherer setting. It is hard to imagine that selfishness and competition between group members would have been more useful than cooperation. As previously pointed out, humans tend to operate not as lone wolves, but as cooperative, colluding confederates. We should also note that modern-day businesses and corporations encourage teamwork, and in fact look to hire individuals that they deem to be "team players."

"Alone we can do so little; together we con do so much"
- Helen Keller

Modernity, with its often large-group, impersonal settings, can sometimes dampen some of our communal urges allowing us to revert to more self-centered behaviors. Our success as a species is a result of how well we communicate, collaborate, and cooperate. The liberal perspective, with its emphasis on the collective, or the common good, is not only more in tune with our evolutionary design, but also with the science of *group dynamics*. We ignore this reality at our own peril.

Perhaps the modern emphasis on the individual (particularly here in the USA) has gone too far. Too often we hear the exhortation that you should not worry or care about what other people think of you, and that you shouldn't concern yourself with other people's opinions of your actions or pursuits. A careful look at our ancient past suggests that in most situations it certainly was beneficial for an individual to care about what the rest of the troupe thought of him. Group consensus and peer pressure was, and still is, very useful for mitigating extreme and errant behaviors. So important was the effective functioning of the human group (band, troupe, clan, etc.) in terms of evolutionary fitness, that we might consider the "group"—not the individual—to be the basic building block of our species.

Having said that, we must understand that it would be foolish and counterproductive for the group to stifle the creativity and initiative of the individual. But these attributes are qualities that nature has endowed the individual with to not only enhance the individual's survivability, but also the group's. Groups are more successful when they value the contribution that the individual makes to the group. And in return, individuals feel good about themselves when they contribute to the group's success. Ideally, individual initiative, creativity, and group cooperation can, and should, go hand-in-hand. Evolution has made it so.

"The autonomy of the individual appears to be complemented and enhanced by the movement of the group; while the effectiveness of the group seems to depend on the freedom of the individual." – Peter Lamborn Wilson (American anarchist author)

CHAPTER 18

PALEO THINKING

"Evolution is an indispensable component of any satisfying explanation of our psychology." - Steven Pinker (Canadian born American psychologist)

Many times we hear people say that they have a "gut feeling," that they "feel it in their heart," or that they can "feel it in their bones." We use these kinds of expressions when we are alluding to feelings that seem to originate from deep within ourselves—some sort of primal yearning or compulsion. *The Paleo Perspective* would argue that we are responding to our genetic inheritance. Let us look at several examples of evolution's possible role in some rather common, everyday human experiences. Please read these with the caveat that these represent general proclivities, and are not true for everybody and all times. They are merely examples of looking at things through a Paleolithic (evolutionary) lens.

- **Water tastes really good when you are thirsty.** This is so that we will provide our body with the necessary fluid it craves. Nature understands that our bodies were designed to be 65% water—and will entice us to keep it that way.

- **Rotting flesh and animal waste smell really bad.** This is to keep us away from some potentially harmful bacteria. Pungent odors are nature's way of telling us to keep a safe distance.

- **Most of us have an aversion to snakes and insects** (and the like). This is to help us avoid some potentially nasty creatures and help steer us away from danger.

- **Most people (especially women) think that babies are cute.** This is nature's way of enticing us to care for them. And so we love to hold them and snuggle with them. We even like how newborns smell.

- **Most people like to listen to and read stories**—particularly stories about people and social situations, both good and bad. People like to talk and to gossip. Nowadays, we find people addicted to their e-mail accounts, their cell phones, and their tablets. We observe people walking across busy intersections glued to their phones. We all want to stay connected. This all makes sense because we are very social creatures and it is advantageous for us to focus on, understand, and be adept at navigating social situations. Communication and socialization are how our species got to the very top of the food chain—it's what we do. Our continued success and survival depend on it. And so we talk, we listen—we socialize.

- **Most people enjoy learning.** Our ability and proclivity to learn is a big part of the success of our species. Nature has made learning pleasurable. It is hard to imagine much human progress if learning was repugnant.

- **Most people find it enjoyable to be near a body of water**—an ocean, lake, or river. In fact, much of the world's population actually lives near a body of water. This is nature's way of enticing us to the shore to take advantage of all the vital resources found in and around water. Besides water to drink, we find lots of sources of protein—fish, shellfish, and other aquatic creatures—at the shore. Rich sources of protein were essential for the expansion of the human brain. Increased intelligence was imperative for the establishment of such a complex, sophisticated, and elaborate human social structure. And so we love the water for good reason.

- **Most males have a fascination with tools and weaponry.** Paleolithic men needed these things to hunt, butcher meat, and protect their clan. Modern men are still instinctively drawn to these instruments. They instinctively understand that these items will help them accomplish the particular tasks that men are many times called upon to perform.

- **Most women seem to have a strong interest in the care of the house or domicile.** They show attentiveness toward keeping the household clean and orderly. This is indicative of a nesting instinct, and probably stems from women being the primary caregiver for the young.

- **Little girls exhibit an inclination to play house and play with dolls.** This is nature's way of preparing them for the possibility and likelihood of becoming a mother.

- **Boredom** is a rather unpleasant human experience. This is nature's way of incentivizing us to be productive.

- **Most people, especially children, have a certain degree of fear of the dark.** In earlier times being instinctively anxious in the deep of night had survival value. Being nervous about nighttime with its many deadly predators encouraged caution.

- **Most people, especially children, enjoy playing various games and sports.** Children's games and sports are instinctually induced practice and training for the kinds of skills one needs for survival. Adolescent play fighting helps prepare the individual for battles to come. Most sports involve some sort of targeting, which involves eye-hand coordination and projectile analysis. These skills come in very handy during a hunt and war. Team sports, in particular, also provide us with the opportunity to express our inherent proclivity be a cooperative, contributing member of a group—a team player.

There are many more examples of how our genetic inheritance contributes to the human experience. Let us also remember that our current genetic design is not the result of any preordained plan. It is instead the result of millions of years of small, random genetic mutations upon which natural selection has operated on.

There may be some aspects of our design that make us wonder why it is so. Some particular features might seem somewhat random and unplanned. There are those who might be inclined to see "the hand of God" at work as we endeavor to understand the human design. Others may be content to take nature at its face value.

Instincts

For your consideration let us add to our present discussion of genetically-shaped behaviors a list of "instincts" generated by Robert Port of Indiana University (April 2000). The list is referred to as "Port's Instinct List."

1. Eat and drink; seek food and water
2. Seek sweet and fatty (nutritious) foods
3. Avoid eating smelly or bitter things
4. Be cautious about novel foods
5. Seek better resources than presently available
6. Seek neighborhoods that are green and flowery
7. Be interested in girls or boys (usually of the opposite sex)
8. Seek sexual contact and excitement; "do the deed"—but in private and not with your children or parents
9. Love children and the cute (female more than male)
10. Compete for resources (wealth, fame, etc.)
11. Pair up; find long-term partners (female more than male)
12. Compete for the best (and, for males, the most) partners
13. Be aggressive, use force to gain advantage (male more than female)
14. Blink; flinch; flee
15. Frown; snarl; attack for advantage (male more than female)
16. Keep yourself alive; ensure your genetic survival
17. Protect your own family

18. Grades of importance: own children > spouse > siblings > clan, etc
19. Obtain and defend resources: jobs, land, property, hunting rights, sex partners, etc. (male more than female)
20. Bond with your group, hang out with your family
21. Try to reduce conflict within your group (female more than male)
22. Compete for leadership in groups
23. Imitate others: beliefs, knowledge and skills
24. Be wary of and dislike "outsiders"
25. Teach youngsters (female more than male)
26. Smile; laugh; mope; frown; strut; cry
27. Celebrate; share gifts (especially food); hug; grieve
28. Display emotions: anger, happiness, disgust, sadness, fear, surprise
29. Be curious about stuff, make sense of things
30. Learn about your surroundings
31. Play with toys (objects, pets, tools, machines, etc)
32. Explore new places, find more efficient methods
33. Be more adventurous—especially when young
34. Learn new words and more effective speech (especially the young)
35. Chitchat; commiserate; boast; scold; make jokes; entertain
36. Organize others' behavior; lead your fellows (male more than female)
37. Tell and listen to stories
38. Make music, make artistic objects

"Be a good animal, true to your animal instincts." - D. H. Lawrence (English novelist, poet, and playwright)

EPILOGUE

The focus of this book has been an analysis of the human condition in modern times. *The Paleo Perspective* argues that in order to make sense of the state of the world we must look at our plight with our Paleolithic inheritance in mind. We must acknowledge our Paleolithic design—both in body and mind. Simply put, we are prehistoric creatures struggling in a modern landscape.

The Paleo Perspective is a cautionary analysis, asking us to reexamine what we consider normal—what we take for granted. It asks us to look at important and consequential things such as sexuality, morality, religion, economics, and politics through a Paleolithic lens.

For example, deviations in sexual behavior are common, and even though a particular sexual behavior is *atypical*, that doesn't mean it is *unnatural*. We should try to react to our fellow human beings' *victimless* sexual proclivities with rationality and tolerance. However, this does not mean that there should be any clemency for *predatory* sexual acts.

Secondly, when we think about morality we should remind ourselves that the forces of evolution, through natural selection, have operated on both the individual and group levels. Man constantly struggles with the conflict between his urge to be selfish and his urge to be a moral and altruistic group member. All examples of altruism and morality are examples of man doing what is best for his fellow man—often without regard for himself.

Furthermore, our concept of morality is closely aligned with the teachings of most modern-day religions. We might think of formalized religion as the codification of our moral values. Whether you believe in a supernatural supreme being or not, we all should adhere to the "Golden

Rule." Indeed, we should look at everything— including economics and politics— through our genetically-fashioned moral lens.

For instance, if we look at the concept of the "corporation" through a moral lens what do we see? We see the corporate entity providing its owners, officers, and stockholders with a construct that makes it easier to avoid taking responsibility for misdeeds. Since the purpose of the corporation is to make money, and because it lacks a "soul" (so to speak), it lacks the capacity to experience guilt, shame, and sin. The corporation gives the owners, officers, and stockholders a degree of deniability and acts as a buffer between them and their Paleolithic conscience.

Similarly, modern man can hide behind the illusion that some benign "invisible hand" guides the machinations of the modern marketplace. This convenient misconception obscures the daily tussle of *intra-species conflict*. It obscures our selfish instinct to maximize profit and grab as large a share of the GDP as possible. It ignores our responsibility to our fellow man—all without guilt. The large, impersonal modern market, like the corporation, makes it easier for man to avoid the Paleolithic emotions that help regulate our interactions with our fellow man.

And what of politics? Americans, in particular, seem infatuated with individual freedom. In doing so, they lose sight of the fact that it is through cooperation and compromise within the group setting that humans excel as a species. Indeed, the common thread throughout our discussion on morality, religion, economics, and politics is the question of *"me* versus *us."* You may recall that the very concept of morality centers on group benefit, not individual gain. And so ultimately, it is all about *"us."* Religious people, agnostics, and atheists alike can all be comforted by the fact that morality is deeply encoded in our DNA—in our ancient Paleolithic design.

We cannot, however, be complacent or feel secure with our innate sense of morality. It is still up to each of us to exercise that morality through our free will. We need to be informed and proactive. Education is particularly critical in a complex and impersonal world. Without education, and without an unflinching sense of self-awareness, we are unwitting hostages of our ancient evolutionary design.

"Knowledge is Power"
— Francis Bacon (English philosopher, statesman, and scientist)

THE FINAL TAKEAWAY

It may be instructive to highlight some of the major concepts espoused in this book. These ideas are essential if one is to fully embrace and employ a *paleo perspective*.

- **In order to be fully self-aware we must acknowledge our evolutionary past** and accept our residual genetic inheritance.

- **Our genetic design hasn't changed in any significant way in the last 200,000 years.** What has changed is the social setting in which we find ourselves. Our ancient genetic design is often ill-suited for modern circumstances.

- **Charles Darwin's theory of evolution is particularly relevant and profoundly consequential.** It has implications for every aspect of human life and is arguably more pertinent than the big bang theory, relativity, or quantum mechanics.

- **Man's evolutionary journey has not been an orderly, linear one.** Man has had a variety of ancestral relatives who became extinct and are not in Homo sapiens' direct line of ascent. Human evolution should not be thought of as some divinely orchestrated journey resulting in a preordained final design.

- **Genes play an important role in human nature.** We have built-in genetic proclivities to help us respond to a variety of environmental circumstances. At times our genes can urge us to be selfish, and at times they can encourage us to be selfless and altruistic. However, genes are not destiny. Ultimately, man has free will.

- **The evolution of the human brain represents a potent, fruitful, and versatile adaptation.** With intelligence comes the ability to alter one's environment, diminishing the need for other (anatomical) varieties of adaptation.

- **Theory of mind is the brain's ability to assign thoughts and intentions to both animate and inanimate entities.** It is essential for the many interactions that take place in highly complex social settings. Theory of mind also facilitates man's rather common belief in supernatural phenomena.

- **Evolution has put pressure on us to survive both as an individual and as a group member.** Our instinct to survive permeates every aspect of our behavior.

- **Man is the most socially complex animal on the planet.** We owe our success to our ability to function effectively in the group format. Man is tribal by nature—a confederate at heart.

- **Any species that is successful enough to fill its eco-niche will experience intra-species conflict.** This conflict is a result of competition for limited resources in the saturated eco-niche. For modern humans, this struggle plays out in the economic, social, and political spheres of society.

- **Morality is ingrained in our DNA.** The evolutionary purpose of morality is to facilitate efficacious group behavior. Morality, empathy, and sympathy are examples of human sentiments that are meant to promote social cooperation.

- **Religiosity is ingrained in our DNA.** Man is hard-wired to entertain religious thoughts and impulses. Religion is a useful social construct that helps promote socially beneficial attitudes and behaviors.

- **Today's capitalistic market economies are complex and impersonal, overwhelming man's moral and ethical instincts.** This has resulted in large-scale income and wealth disparities.

- **Politics is a manifestation of intra-species competition and conflict** that plays out on the local, state, national, and international level.

- **Conservatives and liberals think differently.** In addition to there being a genetic component to political leanings, conservatives tend to focus more on themselves as individuals, and tend to seek more immediate gratification. Liberals tend to focus more on the needs of the collective, and tend to be more willing to postpone gratification in an effort to benefit the collective.

- **The mismatch between our Paleolithic design and modern circumstances is the plight of contemporary man.**

BIBLIOGRAPHY

(Some of the more pertinent references)

Books

Armstrong, Karen. *A History of God*. New York: Random House, 1993.

Bering, Jesse. *The Belief Instinct*. New York: W.W. Norton & Co, 2011.

Dawkins, Richard. *The Selfish Gene*. New York: Oxford University Press, 1989.

Dennett, Daniel. *Breaking the Spell*. New York: Penguin Books, 2006.

Hamer, Dean. *The God Gene*. New York: Random House, 2004.

Krugman, Paul. *The Conscience of a Liberal*. New York: W.W. Norton & Co, 2007.

Lewin, Roger. *Human Evolution*. Malden: Blackwell, 1984.

Murray, Charles. *Coming Apart*. New York: Crown Publishing, 2012.

Nowak, Martin. *Super Cooperators*. New York: Simon & Schuster, 2011.

Skinner, B.F. *Beyond Freedom and Dignity*. New York: Alfred A. Knopf, 1971.

Wilson, Edward. *The Social Conquest of Earth*. New York: Liveright, 2012.

Zimmer, Carl. *Evolution: The Triumph of an Idea*. New York: HarperCollins, 2001.

Journals, Magazines, and Websites

Academia.edu

Cambridge University Press (cambridge.org)

Evolutionary Anthropology journal (onlinelibrary.woley.com)

Journal of Human Evolution (sciencedirect.com)

National Geographic magazine (nationalgeographic.com)

Nature journal (nature.org)

Public Broadcasting Service (pbs.org)

Science magazine (sciencemag.org)

Smithsonian Institute (humanorigens.si.edu)